I0485261

The Extraction of Silver, Copper and Tin from Ores

by James Forrest

with an introduction by Kerby Jackson

This work contains material that was originally published in 1896.

This publication is within the Public Domain.

This edition is reprinted for educational purposes
and in accordance with all applicable Federal Laws.

Introduction Copyright 2015 by Kerby Jackson

Introduction

It has been over a century since James Forrest released his important publication "The Extraction of Silver, Copper and Tin". First released in 1896, this work has been unavailable to the mining community since those days, with the exception of expensive original collector's copies and poorly produced digital editions.

It has often been said that *"gold is where you find it"*, but even beginning prospectors understand that their chances for finding something of value in the earth or in the streams of the Golden West are dramatically increased by going back to those places where gold and other minerals were once mined by our forerunners. Despite this, much of the contemporary information on local mining history that is currently available is mostly a result of mere local folklore and persistent rumors of major strikes, the details and facts of which, have long been distorted. Long gone are the old timers and with them, the days of first hand knowledge of the mines of the area and how they operated. Also long gone are most of their notes, their assay reports, their mine maps and personal scrapbooks, along with most of the surveys and reports that were performed for them by private and government geologists. Even published books such as this one are often retired to the local landfill or backyard burn pile by the descendents of those old timers and disappear at an alarming rate. Despite the fact that we live in the so-called "Information Age" where information is supposedly only the push of a button on a keyboard away, true insight into mining properties remains illusive and hard to come by, even to those of us who seek out this sort of information as if our lives depend upon it. Without this type of information readily available to the average independent miner, there is little hope that our metal mining industry will ever recover.

This important volume and others like it, are being presented in their entirety again, in the hope that the average prospector will no longer stumble through the overgrown hills and the tailing strewn creeks without being well informed enough to have a chance to succeed at his ventures.

Kerby Jackson
Josephine County, Oregon
January 2015

Mav. 18, 1939 WITH(1)/AE

THE INSTITUTION OF CIVIL ENGINEERS.

SECT. I.—MINUTES OF PROCEEDINGS.

17 March, 1896.

Sir BENJAMIN BAKER, K.C.M.G., LL.D., F.R.S., President,
in the Chair.

(*Paper No. 2898.*)

"The Lixiviation of Silver Ores."[1]

By JOHN HENRY CLEMES, Assoc. M. Inst. C.E.

IN most silver-producing districts, ores occur containing insufficient lead or copper to be submitted to the smelting process, and too impure to be dealt with by the wet-milling processes. In many instances the yield of silver from such ores is too low to permit their transport to metallurgical works where they can be dealt with economically, and it becomes necessary to treat them by such methods as can effect the purpose cheaply under the prevailing economic conditions. The presence in an ore of considerable amounts of sulphides, sulpharsenides, or sulphantimonides of the base metals, renders it irreducible by the amalgamation process. These compounds, the presence of even small quantities of which suffices to remove an ore from the free-milling category, are, for the most part, limited to iron pyrites, copper pyrites, copper glance, galena, blende, mispickel, and the various fahlores. Other varieties of these compounds occur not infrequently, but the cause of the so-called "rebelliousness" of an argentiferous ore is generally the presence of one or more of these minerals.

The almost universal method of winning the silver from such ores, was to first submit them to a chloridizing roasting, and then to treat the roasted product by one of the amalgamation processes. With certain types of these ores, the results obtained by this procedure were unsatisfactory, either the loss of silver incurred in the reduction, or the actual cost of the reduction was considerable. The use of sodium hyposulphite as a solvent of silver chloride in ores was proposed by the late Dr. Percy in 1848, and the first

---

[1] The discussion upon this communication was taken in conjunction with the two following Papers.

trials on a working scale were carried out by Von Patera ten years later. The development of the process was commenced in Sonora, Mexico, by Mr. O. Hofmann; and two or three lixiviation plants, designed and constructed by him, were in successful operation by the year 1870. The attention of mine-owners who had to deal with refractory ores was then directed to the process, and its use began to increase in Sonora, in the other north-western States, Chihuahua and Sinaloa, and in the Federal district of Baja California. It proved well suited for the treatment of certain classes of ore of frequent occurrence in those districts.

The extraction of silver by the lixiviation, or Patera process, is effected by converting the silver salts contained in the ores into silver chloride, which is dissolved in a solution of calcium hyposulphite, from which the silver is precipitated by an alkaline sulphide. The three stages by which these reactions are carried out are, roasting the ore in admixture with common salt, leaching the roasted ore with a dilute solution of calcium hyposulphite, and precipitating the silver sulphide from this solution by the addition of calcium sulphide. The precipitation of the silver is attended by the regeneration of the hyposulphite solution; the addition of calcium sulphide to the solution of silver chloride in calcium hyposulphite, in a proper proportion, precipitating silver sulphide and free sulphur, with the simultaneous formation of calcium hyposulphite. After the precipitate has settled, the supernatant hyposulphite is drawn off and pumped back to do duty as fresh solvent, and, under favourable circumstances, increases in strength and in volume. It is, however, subject to deterioration, partly from impurities and partly from its own instability; ores free from substances noxious to the solution are comparatively scarce, and the decomposition of the solution, from its slow oxidation, is always in progress. The remedies are, however, in most cases, of easy application, and the process is capable of dealing with a very wide range of silver ores.

With the exception of those containing a considerable percentage of galena, all silver-bearing ores, associated with these minerals, are effectively treated by the lixiviation process. Ores containing a small percentage of copper are treated with excellent results, and the rapidity with which the use of the process extended in Sonora is partly due to the fact that nearly all the silver ores found in that State, argentiferous galena excepted, are accompanied by some mineral of copper.

As compared with the means of extracting silver from its ores by roasting and amalgamating, the Patera process offers the advantages of employing a much cheaper solvent than mercury, and of not requiring, in the treatment of the roasted pulp, any expensive machinery; power for stirring and grinding the roasted pulp is not needed, nor have a number of the working parts of the apparatus to be frequently replaced. The cost of working by either process is much influenced by local conditions, but, in general, if the nature of the ore is such as to require a chloridizing roasting, its contained silver can be more cheaply extracted by a solution of an hyposulphite than by means of mercury.

PLANT EMPLOYED IN THE PROCESS.

The plant for the lixiviation process comprises machinery for breaking and pulverizing the ores; roasting-furnaces; leaching- and precipitating-vats; and apparatus for treating the final product, or the silver precipitate.

The stamp-mill, roasting-furnaces, cooling-floors, and leaching works are set terrace fashion on the hill-side, a suitable slope for the purpose being one of 25° to 30°. The furnaces are placed with their long axes about at right angles to the contours of the hill, and at such an elevation that a railway may connect their feed-hoppers with the pulp-bins in the mill. The level space for the cooling-floor is set out a few feet below the points of discharge of the furnaces, and a narrow-gauge line from it passes over the tops of the ore-vats. Following the practice of mills in which roasting and pan amalgamation are used, the various departments of a lixiviation mill are generally collected under one roof. Ease of supervision is gained by the adoption of this plan, but in the event of a serious fire there is much risk of losing the whole plant. The bulk of the structure is of wood, even the roof covering is of wooden shingles, heavy import duties, and heavy rates of freight, placing the use of corrugated iron out of the question, and the climate is extremely dry. This risk renders it preferable to detach the furnace and leaching departments from the mill proper, and, in large establishments, it appears advisable to detach the leaching works from the roasting-sheds. At the Yedras mines the reduction works had to be arranged *en bloc*; fire-hydrants were placed at different points of the works, and were connected with water-tanks situated several hundred feet above the mill.

Grinding Machinery.—In northern Mexico the ores are almost always ground by a dry stamp-mill. In the 40-stamp mill built at the Yedras mines, the ore—much of which arrives partially broken from the picking-floors—is fed into two stone-breakers in the upper part of the mill and from them drops into the ore-bins. From sliding-doors in the bottoms of these bins the ore is run into wagons and tipped into the dry kilns, and from the cast-iron spouts at the base of the latter it is conveyed to the self-feeders at the back of the batteries. Horizontal conveyors in front of the batteries transfer the pulp, or pulverized ore, to the inclined elevators, which raise the pulp to the hoppers in which it is stored. From these storage-hoppers the pulp is conveyed to the small hoppers over the roasting-furnaces. With the exception of wagon-filling from inclined spouts, tramming and tipping, the process of converting the stone into pulp of the required degree of fineness is accomplished entirely by mechanical means. The crushed product required by the lixiviation process is fine meal, as free as practicable from impalpable powder or slime.

The Yedras mines are situated about five days' journey from the coast, in a spur of the main Cordillera, at an altitude of more than 7,000 feet above the sea-level, and all freight has to be carried to them on pack animals. Heavy parts of the machinery had therefore to be divided into pieces not exceeding 350 lbs. in weight; a few parts of the apparatus which could not be so treated were attached to poles and carried over the mountains by large relays of men. The battery-posts were of the most thoroughly seasoned timber available. The feed side and front of the batteries below the line of the braces were unobstructed, so that the automatic feeders, the mortars, and the pulp-conveyors were readily accessible. Instances of the employment of the double mill for dry stamping are to be met with; in this type of mill two parallel lines of battery-posts are set facing each other, and united by heavy tie-beams, braced one against the other. This mode of construction gives a very strong and steady battery-frame; but, for dry-crushing purposes, is attended with disadvantages. In the one case known to the Author of the adoption of this type of mill, the conveyor-system used for transferring the pulp was complicated, and had the further disadvantage of being in part actuated by gearing. The presence of grit and dust in these positions renders the use of gearing very objectionable; another objection to this design of battery-frame for dry crushing is the want of sufficient space for the dry kilns.

The effecting of pulverization with a minimum production of slime calls for special attention in leaching-mills, since the rate of filtration in the vats is dependent on the physical character of the ore-pulp. One of the weakest points in the older forms of the stamp-mill, when used to prepare ores for the leaching process, was the reduction of a large amount of the ore to an impalpable powder. The changes introduced to increase the output have had the further effect of improving its quality. Less dust is present in the pulp produced by the modern forms of stamp-mill than in that produced by the older forms, because the action of the former is more rapid than that of the latter. The ore remains a shorter time in the mortars, and a smaller amount of the crushed product is subjected to injurious pulverization. The mortars are now furnished with screens on both their long sides, while in the older practice the screens were placed on the front only. The configuration of the mortar itself is better adapted for rapid discharge, the dies are set higher, the machine is made stronger and stiffer, so that heavier stamps and higher speeds can be used. The force with which the partially-crushed product is driven from the anvils, and the facilities of escape offered to that part of it which is sufficiently crushed, are increased by these changes. Much less, therefore, rolls back on the anvils to be pounded afresh by the descending hammers. Other changes operate to diminish the production of over-crushed pulp, such as the more extended use of rock-breakers, the mineral being served to the batteries in a much smaller form; and the equally extended use of automatic feeders. As a consequence, the stratum of ore interposed between the stamp and its anvil is kept thinner in the modern types of mill, and the wave-like motion of the pulp against the screens is better maintained—a very important matter in a dry mill. By using a heavy stamp, the dash of the pulp against the screens, and, therefore, its chance of escape from further comminution, is increased; but the shattering action on the ore, as compared with that produced by a lighter stamp, is also increased. The advantages seem to lie on the side of a stamp weighing between 950 lbs. and 1,000 lbs., when, as is nearly always the case, the preliminary breaking of the ore is limited to a passage through a rock-breaker set to crush to $1\frac{1}{2}$ inch to $1\frac{3}{4}$ inch. Whether the practice of breaking the ore to a much smaller size before feeding it into the batteries, and of using light stamps for the final pulverization, would sufficiently better the physical condition of the pulp to justify the increased complexity of the mill is a matter worthy of careful investigation. Experiments with various types of ore have repeatedly shown that finer breaking in the rock-breakers is followed by increased duty from the mill; thus the preliminary breaking of an ore to the size of coarse gravel, as against breaking to the size due to passage through a rock-breaker with jaws set $1\frac{3}{4}$ inch apart, will increase the output from the batteries some 15 per cent. or 20 per cent., the other conditions being the same. The introduction of the multiple-jaw breaker has greatly facilitated the preliminary reduction of ores to fine sizes. The lessening of the amount of ore ground to an impalpable powder is effected also by the following method. A screen of much coarser mesh than that needed for effective chlorination is placed on the mortar, and the resulting pulp is passed through a revolving screen of the required mesh, the coarse particles being returned to the mortar. This simple modification of the usual practice will frequently effect a great improvement in the leaching qualities of the pulp obtained in crushing brittle dust-forming ores. In a test by the Author with an ore of this type, the rate of filtration in the leaching-vats was nearly doubled by the adoption of this expedient. The 24-mesh sieves used on the batteries were replaced by heavy jig-screens having only 6 meshes per lineal inch, the coarse pulp was elevated a few feet above the cam-shaft floor, and passed through a revolving 24-mesh sieve. The tail division of the housing which surrounded this revolving screen terminated in a pointed bin from which the rejected gravel was run into a wagon and returned to the batteries. The evil of over-grinding is sure to be augmented if the means employed for drying the ores are inadequate to the requirements of the mill; the smallest trace of moisture in the ore fed to the batteries suffices to retard the discharge of the pulp and to increase the amount reduced to an impalpable powder. It is an advisable precaution to furnish leaching-mills with dry kilns of a capacity somewhat in excess of their normal requirements. An efficient apparatus for the suction and removal of dust from the upper parts of the mortars should not be omitted, since the most elaborate precautions of tight housings, or packing around the stamps' stems, fail to prevent the escape of ore-dust into the mill, with the attendant ill effects on the health of the attendants and to the working-surfaces of the machinery.

In one or two instances in which a somewhat coarse grade of crushing proved sufficient, roll crushers have been used in Sonora for grinding ores. At Promontorios the weathered ores, proceeding from the upper levels of the mines, were, for a long period, crushed through a pair of rolls 30 inches in diameter by 11 inches wide, and provided with screens having 12 meshes to the lineal inch. These screens were heavy, for use in connection with the jigging plant used at these mines, and were therefore equivalent to a somewhat finer sieve of the type ordinarily used with stamp batteries. As soon, however, as the unaltered sulphide ores of the points below the drainage level were reached, a greater degree of fineness of crushing became necessary, and for this the single pair of rolls employed proved so inadequate that recourse was had to stamping. For lixiviation, the physical character of the pulp produced by the roll crusher is superior to that produced by the stamp, being more granular, and much more free from dust, and the percolation of liquids through it is therefore more rapid.

In the few modern mills of this type which have been built for leaching works, the crushers are more strongly built, and better adapted to the work required of them, than was the case in the older machines; the rolls are driven at a higher speed, and the reduction of the ores in one pair of rolls, from the size produced by a coarse rock-breaker to the condition of fine meal, is abandoned. Not only is the preliminary breaking carried further than formerly, but the actual crushing is effected through two pairs of rolls. The results obtained have, however, not been entirely satisfactory; in one or two excellently built mills the crushers, when submitted to the test of continuous hard work, have failed to supply the furnaces with the amount of fine meal required of them. The roll crusher, although without a rival as a reducer of ore to small gravel size, has but a limited capacity when used as a fine pulverizer. So that if roll crushers are to be successfully employed for the production of fine meal, for such purposes as those of the leaching process, the principle of graduated reduction in a series of machines must be more thoroughly applied. In view of the wear and tear of the rolls and sieves, it is doubtful whether any cheapening of the operation of crushing would be brought about by substituting rolls for stamps; the advantage to the lixiviation process would be the much improved condition of the crushed ore, both for the chloridizing furnaces and for the filtering vats.

Roasting-Furnaces. — The long reverberatory furnace is, in Northern Mexico, used more generally in connection with the lixiviation process than any other type of roaster. The tendency towards the adoption of mechanical furnaces has, however, increased of late years; the scarcity of labour in the mountainous districts offers a strong incentive to the employment of these labour-saving devices at the mines, the ores of which are susceptible of treatment by their agency. There are, however, in a number of instances, ores unsuitable for this mode of treatment, and the owner has no choice but to use the labour-consuming reverberatory furnace.

The length of the reverberatory furnace to be adopted depends on the amount of sulphur contained in the ore, and, in most cases, varies between 40 feet and 60 feet. This space is divided into several hearths, usually 10 feet to 12 feet long; a usual type of furnace for an ore moderately charged with sulphides having four hearths each 11 feet long, and for an ore more highly charged with sulphides five hearths each 12 feet long. With ores rich in pyrites a length corresponding to six 12-foot hearths is sometimes employed; the latter may be regarded as the maximum length, and no useful purpose is served by increasing it. The hearth floors are horizontal, that nearest the fire-bridge being the lowest, and each successive one is set about 3 inches higher than that contiguous to it. Occasionally these steps are omitted, the whole hearth lying in one horizontal plane; the omission, however, is not to be recommended as the steps serve to lessen the mixing of the various ore charges. The furnaces are about 10 feet in width when provided with working doors on one side only, and 12 feet to 12½ feet when furnished with working doors on both sides; the former are much more frequently found in use than the latter.

One of eight furnaces built by the Author at the Yedras mines is shown in Figs. 1, Plate 2. The only ground available for these furnaces was considerably steeper than desirable, and, partly with the view of lessening the amount of excavation needed at the inner ends of the furnaces, the outer ends were set over arched vaults. Into these vaults, the floors of which were highly inclined, the roasted ore charges were dropped in a red-hot condition, and, after cooling, were run into wagons and transferred to various points of the cooling-floor. The vaults were freely supplied with air, and the fumes from the glowing charges were carried off by flues passing under the centre of each furnace and communicating with the dust-chambers. In many ways this mode of discharge proved superior to that of dropping the red-hot charges into wagons and dumping on the cooling-floor; the entry into the works of the dense clouds of chlorine fumes, and of roasted ore dust, which are given off during the discharge, is suppressed, the amount of chlorination which ensues after dropping the charge—the so-called "heap chlorination"— is increased, and the furnace is much more rapidly emptied, as no time is lost in withdrawing one wagon and bringing up another. The further important advantage is obtained that much of the unhealthy cooling-floor work, such as that of sprinkling water on hot ore, is obviated. Where wide furnaces with doors on each side are adopted, this system may with advantage be carried a stage further, by discharging from one side of the furnace for a certain number of hours into one compartment of the vaults, and then from the other side into another compartment; by these means the pulp can be almost entirely cooled before removal. One reason for the almost exclusive use of the narrow furnace is the fact that the few wider furnaces introduced were still of small dimensions, so that the increase in the number of men required for their manipulations was out of proportion to the increased volume of the ore charge, and the saving in fuel was largely off-set by an augmented labour cost. Great reductions in the cost of roasting copper ores have, of late years, been effected in the western part of the United States by the employment of reverberatory furnaces of large dimensions.[1]

In many of the mountainous districts of Northern Mexico the labour supply is very irregular, and work has to be interrupted and resumed in accordance with the fluctuations of this supply. The roasting-furnaces are therefore occasionally shut down and again fired up; since this process of successive cooling and heating, which greatly shortens the life of the arches of the narrow furnaces, is destructive to those of the wide type, the retention, under such conditions, of the smaller furnace seems advisable. In the more settled parts of the country, however, where this influence does not apply, the use of much wider furnaces is practicable, and would materially diminish the cost of roasting. The walls of the furnace up to about 1 foot below the line of the hearths are of brick, or rubble, set in lime mortar; and above that point of brick set in clay mortar; they are thick in order to keep the exterior as cool as possible, the climate renders this precaution indispensable. Fire-brick is used only in the lining of the fire-box and fire-bridge, and the short arch covering them. At Promontorios the fire-box walls and bridge were made of cut blocks of a refractory stone found in the vicinity. The space under the hearths is filled with dry sand or dry ore-tailings, or with fine screened gravel; the use of rubble or débris is avoided as tending to produce uneven settlement. The hearth is paved with red bricks laid on their long edges, and the outer row of each step with bricks laid on end; the hardest bricks, those from the arches of the burning kilns, are selected for the hearths. The bricks are made on sand floors; those used for the hearths and the main arches have therefore to be rubbed down to remove roughness. At the Rosario mines a brick-making machine was comprised in the constructional plant, and its cost was returned many times over in the increased expedition of the work and the better quality of

---

[1] "Modern Copper Smelting," by Edward Dyer Peters, pp. 445–470. New York, 1895.

the bricks furnished.   In the main arches, tiles equal in area to two bricks are used, and the number of joints is thereby halved; the arches so made last as long as those made of ordinary brick. Immediately over the wide low fire-bridge a few holes are left in the arch, through which air, heated by passage through the hollow walls and over the arch of the fire-box, is admitted; for the same purpose apertures communicating with the interior of the first hearth are left in the fire-bridge itself; two sliding doors in the outer wall serve to control the amount of air admitted at these points.   As the working doors are never closed whilst the furnace is at work, the atmosphere in its rear and middle parts is highly oxidizing, but that of the first hearth may, unless the above precaution is taken, be at times the reverse of oxidizing; the pitch pine so largely used as fuel in the Sierra, as well as many of the hard woods of the coast, give off, immediately after firing, great volumes of smoky reducing flame.

The dust-chambers between the furnaces and the chimney are frequently too small to be effective; in some mills they are omitted, a short stretch of flue being relied on to collect the ore dust.   Where mechanical roasters are used, the furnace gases are passed through a long chamber divided horizontally and vertically by thin partitions, so as to form a serpentine, causing the deposition of dust to take place at the bends.   At the Yedras mines the furnace gases were passed through such a chamber, and had further to traverse several hundred feet of spacious flue before reaching the chimney.   It would undoubtedly be an improvement to construct the entire length of flue in short stretches, so that the whole could be cleaned from the outside by the hoe.   The incessant changes in the direction of the gases would make such a flue a much more effective collector of dust than the ordinary straight one, and a somewhat smaller length might therefore be adopted; the efficiency of this zigzag flue would be increased if several of the sections were made about three times as wide as the average width of the flue, so that the fumes would have to traverse, at certain intervals, zones of slowly-moving current.   The height needed for draught, and for discharging the noxious fumes at a safe distance, is generally obtained by running the flue up the side hill, the chimney being merely built high enough to have its top clear of surface currents; the stacks at Promontorios and at Yedras were 6½ feet square inside and 54 feet high.

The allotting of ample space for the cooling-floor—the paved area on which the roasted ore is spread and moistened—cannot be too highly recommended.   Where it is either too small, or improperly located with reference to the rest of the works, the operations conducted on it are expensive and tedious. Even where the area assigned to the cooling-floor is ample, and where the facilities for conveying the roasted pulp to and from it are good, these operations call for a large expenditure of· manual labour, and, with many types of ore, are unwholesome.

The Leaching Plant.—The operations of leaching are now invariably carried out in circular wooden vats. In the early days of the use of the process rectangular wooden tanks were employed, but the difficulty of making and keeping this form of receptacle perfectly tight led to its abandonment. Three types of vat are made: the ore- or leaching-vats, the precipitating-vats, and the storage-vats. The leaching-vats have to serve as filters, and are provided with false bottoms.

Under normal circumstances, the useful sizes of vats lie between a diameter of 15 feet for use in small mills, and of 25 feet for use in large mills, those with a diameter midway between these limits being very well suited for use in mills capable of treating 50 tons to 60 tons of ore per day. In the event of the construction of very large works these diameters might well be greatly exceeded. The depth of the ore-vats depends, to a certain extent, on the character of the ore, and the degree of fineness to which it is ground. Few ores, even argillaceous ores, have a tendency to form slime, and so hinder filtration, after a thorough roasting, those containing lime being perhaps the worst from this point of view. The other cause of slowness of filtration, the presence of dust in the pulp, is of more frequent occurrence ; the metallic constituents of many ores are extremely brittle, and are, during pulverization, reduced to an impalpable powder, even when comparatively coarse screens are used. To guard against possible difficulties in the way of filtration, the ore-vats in the older establishments were made with a depth above the false bottom of only $3\frac{1}{2}$ feet to $4\frac{1}{2}$ feet, but the tendency has always been to increase this depth, and vats with a depth of $5\frac{1}{2}$ feet to $6\frac{1}{2}$ feet are now commonly used. At the same time the height at which these vats are set over their recipients has been increased from about 2 feet to as much as 6 feet, so that the pressure on the surface of the filter is considerably increased. The usual method of discharging the spent ore is to shovel it over the sides of the vats into V-shaped launders. Water is at most of the mines very scarce, and has to be carefully husbanded ; these launders are therefore set at a gradient so steep that a very small stream suffices to carry off the tailings. In some cases even this amount of water cannot be spared and the tailings are shovelled into wagons. Some of the vats built for use in the Russel process are 7 feet and more in depth, and the tailings are removed by sluicing—the best method of removing them if the necessary water is obtainable; the false bottoms of the vats are set horizontal. In this process, ejectors, or other suction appliances, are used to increase the pressure on the surface of the filters, and the use of deep vats is recommended under all circumstances; where sufficient water for sluicing does not exist, the removal of tailings in buckets hoisted by mechanical means is to be recommended; the latter mode is in use at some of the works in which gold ores are treated by the cyanide process. A leaching-vat has been devised by Mr. O. Hofmann in which the false bottom has the form of a blunt funnel, and the spent ores are sluiced through the centre of the vat; the object of this mode of construction is to lessen the quantity of water needed for sluicing. The false bottoms are composed of narrow wooden slats attached to the bottoms of the vats of the perforated boards which rest on these slats, and of the cloth through which the material is filtered. The holes in the perforated boards are about 1¼ inch in diameter, and 3 inches apart from centre to centre. Two holes are bored in the vat below the false bottom, one to serve for the drawing off of the solution, the other for the escape of the air which lodges in this space. Around the edge of the vat some wooden segments are laid over the filter cloth and tightened by wooden wedges inserted at two or three of their joints. The cloth is somewhat larger in diameter than the vat, and the surplus is pressed between the segments and the edge of the vat. Filter-beds consisting of broken slag or stone surmounted by a thin layer of gravel, such as tailings from jigs, have been sometimes used. Such bottoms need be disturbed but rarely, and offer other advantages; but they have the disadvantage of taking up much more of the useful depth of the vat than the board-and cloth-filter; where the removal of tailings is to be done by sluicing, the use of such beds is of course inadmissible. A convenient size of vat for the precipitation of the silver sulphide is between 12 feet and 13 feet in diameter by 10 feet deep.

The timber commonly employed on the coast belt for the construction of vats is a variety of cypress named sabina; it is very durable, and admirably adapted for such purposes. Cedar is readily and cheaply obtainable at many points, and is equally suitable. Californian redwood is also suitable for these purposes; at the Rosario mines, which were connected with the seaboard by a wagon road, the vat staves and vat bottoms of this material, prepared by machinery, were shipped from San Francisco. In works situated in the mountains, the vats have to be made of the pine of the locality, the only timber available.

In modern lixiviation works the vats are usually placed on timber trestles resting on low walls. At mines where long timber is unobtainable, the supporting walls are raised a few feet above the ground, and the joists which receive the vats are placed directly on them; in such localities each line of vats is covered by a low shed supported on posts of native hard woods. At the pioneer lixiviation establishments in Sonora, the tops of the ore-vats stood only 2 inches or 3 inches above the surface of the ground, and the roasted ore was wheeled to their edge and tipped into them. This mode of erection, which is open to several objections, has long been discontinued. The natural and the most convenient arrangement of the plant is to set the vats in tiers, Figs. 2, Plate 2, the ore- or leaching-vats at the top, and the precipitating-vats and storage-vats following in order. Among the more important of the subsidiary portions of the plant are the means employed for elevating and conveying the solution of calcium hyposulphite. The solution attacks brass and gun-metal, but has no effect on hard antimonial lead. On cast-iron surfaces, with which it is merely brought in contact, as in pipes of that metal, the attack is so very slow that it may, in practice, be disregarded. Under certain conditions, however, the rate of attack becomes noticeable; for instance, the friction between the piston and cylinder of a double-action pump renders the rate of attack sufficiently rapid to make the use of such apparatus inadvisable. With this condition modified, as in the case of a cast-iron plunger in rubbing contact with a soft packing, the rate of attack is again practically unimportant. But another characteristic of the solution has to be taken into account; on all iron surfaces over which the solution moves slowly, or is sometimes permitted to rest, a firmly adhering scale of gypsum is formed, and the rate of deposition is sufficiently rapid to enforce frequent dismantling and cleansing of the apparatus. Where, however, the conditions for such deposition are unfavourable, for instance, where the current is rapid and long horizontal waterways do not occur, the difficulties from this source are reduced to a minimum; for this reason the centrifugal pump, if wholly of cast-iron, is an efficient apparatus for the raising of this liquid. In the erection of the pump, the discharge-pipe is set vertical, and has no turns or elbows, but is united by a flange to the bottom of a large wooden trough, and the pump is set low so that only a short suction-pipe is required; with these precautions pumps of this type work well for many years, and give but little trouble. If very deep vats are used, it may be necessary to use two centrifugal pumps, one delivering to the other. For distributing the solution, the use of iron pipes is altogether eschewed, and wooden launders used; the connection between the vat and the launders is made by stout rubber hose, $1\frac{1}{2}$ inch in diameter, attached to short lead tubes inserted in the bottoms. No metallic stopcocks or valves are used; the flow of solution is sometimes regulated by screw clamps placed over the hose, but oftener by tapering wooden plugs inserted in the extremities. Gypsum is deposited to a small extent on wooden surfaces, but is easily detachable from them; the only part of the plant where its presence is of any consequence is in the space beneath the false bottoms of the leaching-vats, and this is cleaned every two or three months. The holes of the false bottoms are always burnt out after being drilled, and no deposit forms in them. The filter-cloths are frequently washed, and last several months. For the manufacture of the calcium sulphide solution a species of churn, with cast-iron bottom and sheet-iron sides, is used, a convenient size being 6 feet in diameter by 9 feet deep; a discarded amalgamation-pan is sometimes used for this purpose. This solution has no effect on iron, and is conveyed in ordinary gas-pipe; it must be stored in iron vessels.

The Manipulation of the Process.

Chloridizing-Roasting of the Ore.—The results obtained by the use of the process depend very largely on the way in which the operation of roasting is carried out. When it is well conducted the close and economical extraction of the silver, by the hydro-metallurgical operation which follows, is practically assured. On the first employment of the lixiviation process, the roasting of the ore was accomplished in short single-hearth reverberatory furnaces, which did not greatly differ in form from the old types of calciner used with copper ores. Their use has now been entirely abandoned in favour of the long multiple-hearth reverberatory furnace, or, of furnaces in which the ore is stirred by machinery. A number of advantages attended the change from the single-hearth to the multiple-hearth furnace; the work is more expeditiously and cheaply accomplished; the roasting operation is a continuous instead of an intermittent one, because the necessity of allowing an interval for cooling between the withdrawal of a roasted charge and the admission of a raw one is dispensed with.

The mode of treating an ore in a long reverberatory furnace may be regarded as typical of that employed in any form of furnace, and the same remark applies to the various chemical reactions which take place during the treatment. For this reason the operation of roasting is here described as carried out in this form of furnace. For ores which present no abnormal difficulty, and for which the process has been largely used, such as those in which iron pyrites, with a small admixture of copper pyrites, fahl ore, or galena, are the chief metallic components, and the matrix is for the most part quartz, the manipulations may be briefly described as follows : A charge of ore, crushed to a fineness due to a passage through a sieve, with twenty-four to thirty holes per lineal inch, and to which between 3 per cent. and 6 per cent. of common salt has been added, is run from the hopper into the hearth furthest removed from the fire, say the fifth, and is at once spread out with the rabble. The weight taken for a charge may be 16 lbs. to 18 lbs, for each square foot of the area of the hearth. With an ore of this type the charge will not remain in each hearth longer than one or one and a half hour ; in the fifth hearth it will receive scarcely any raking or stirring, since at this stage the ore runs like quicksand on the least touch with the rake or rabble. At the end of this period the charge is advanced to the contiguous hearth, all the charges in the furnace being simultaneously advanced one stage. The arches of the furnace, with the exception of the first, are at all points in close proximity to the sole of the furnace, so that the surface of the ore is exposed to a stream of red-hot air charged with the products of the combustion of sulphur and of the decomposition of sodium chloride, and the draught is so regulated that this current of hot gases rolls slowly along the surface of the charges. Under these conditions a portion of the sulphur of the pyrites ignites soon after the removal of the charge to the fourth hearth, and volatilizes as sulphurous acid ; the mass is stirred as often as the blue flame of the burning sulphur disappears, in order to expose fresh surfaces to the action of the air. This combustion of one part of the sulphur of the pyrites affords sufficient heat to bring the charge to the incandescent condition, in which it is passed on to the third hearth, where the circumstances are highly favourable to the formation of sulphates, a very important part of the operation. The charge itself is in a condition of readiness. It is in a dull-red glow, and at the same time has but little tendency to sinter or form hard balls through the partial fusion of its sulphides, the decomposition of the latter being sufficiently advanced to remove this tendency. The atmosphere in the interior of the hearth is at the right temperature and highly oxidizing, as the amount of air admitted at all stages of the process is much in excess of that needed for the combustion of the sulphur. The gas evolved from the ore charge at this stage is sulphurous acid, intermixed with a large amount of chlorine and some sulphuric anhyride proceeding from the advanced charges in the two lower hearths. The presence of the two last-named gases increases and quickens the reactions in the third hearth. In this hearth the charge becomes coherent and begins to swell. About an hour after its entry signs of the beginning of chlorination are perceptible. Chemical tests show that this action has been in progress from the first entry of the ore into the furnace, but up to this point it has not been manifest to the eye. The reactions in the third hearth are therefore the combustion of sulphur with evolution of sulphurous acid, the formation of sulphates, and incipient chlorination. In the second hearth the charge receives much more stirring than hitherto, and its chlorination is much advanced. In the finishing-hearth the temperature, immediately after the dropping of a spent charge, is that of a dull cherry-redness. Since this temperature suffices, with most classes of ore, to bring about the mutual decomposition of the sodium chloride and the various sulphates formed in the earlier stages of the roasting, the charge, on being exposed to it, presents all the phenomena which attend active chlorination. It greatly increases in volume, becomes slightly pasty, and gives off, at all surfaces exposed to the air, dense volumes of chlorine intermixed with small amounts of hydrochloric- and sulphuric-acid gases. On the first use of the process in Sonora, steam was admitted over the fire-bridge for the purpose of increasing the formation of hydrochloric-acid gas, but no improved results followed from this practice, and it was abandoned. With the same object in view, green wood was sometimes used as fuel, but its use also ceased. During the period of very active chlorination the fire on the grate is kept low, and air is admitted to the full extent of the apertures provided for the purpose in the arch and the fire-bridge; that the interior of the distant hearths may not be chilled by this large introduction of air, the advantage of previously heating it, by passing it through the walls and over the arch of the fire-box, is manifest. The usual practice of admitting air through the doorways of the ash-pit and the fire-box is very wasteful of fuel. The chemical reactions in progress at this stage are accompanied by a great evolution of heat, by which, supplemented in other parts of the furnace by that arising from the combustion of sulphur, the temperature over the great length of the furnace is practically maintained. When, by the lessened evolution of chlorine, the change of colour of the ore-pulp, and the general toning down of the vigour of the reactions, the operation is judged to be nearing completion, the supply of air is reduced and a few sticks of wood are thrown on the fire-bars. Samples are frequently drawn from the charge, cooled quickly on an iron plate, and carefully examined. Since these samples are daily tested in the laboratory the physical characteristics of properly roasted pulp soon become familiar, and the exact moment at which any charge is dropped is determined from the simple inspection of these samples.

Many of the difficulties which surrounded the roasting of several classes of ore in the single-hearth furnace have been removed by the use of the long furnace. Ores, for instance, containing a large quantity of easily fusible sulphides could not be treated in the short furnace without the formation of balls from the sintering, or partial fusion, of these substances. And, since the silver contained in these balls could not be extracted by the solutions, it became necessary, either to break them up on the hearth of the furnace with heavy tools, or to pass the whole of the roasted ore through sieves, to return the coarse part to the grinding machinery, and to re-roast it with raw ore; neither of the remedies was efficient, and both were expensive. Difficulties from this source are rare with the long furnace, the sulphides lose much of their sulphur in the cooler hearths, and show little or no disposition to sinter on reaching the hotter hearths. The use of the long furnace is especially advantageous in the treatment of ores rich in arsenic, the sulphide of which volatilizes at a low temperature, and the arsenious and sulphurous acids resulting from its dissolution are carried off by the draught. If, however, during the first stages of the roasting the temperature is increased, this desired volatilization is checked, the formation of arsenates takes its place, and the presence of these salts retards the roasting in its later stages. Every effort is therefore made to eliminate all the arsenic possible in the cold stages of the roasting, and the chief means to this end, the maintenance of a low temperature, is automatically provided in the long furnace; occasional carelessness in the matter of firing cannot, owing to the distance of the final hearths from the fire-grate, do away with this condition.

The gangue, or matrix, associated with the ore, plays an important part in the roasting process. Owing to its being less brittle than the ore proper, it issues from the grinding machinery in a much coarser form than the latter, varying in size—according to the sieves employed—between coarse sand and fine meal. The presence in the ore pulp of large amounts of this coarsely-crushed material renders it porous, open, and permeable to the furnace gases.   This fact is often disregarded in small establishments, where the desire to keep the yield of the ore in silver at the highest attainable point causes the work on the breaking floors to be carried much too far and the gangue as far as possible eliminated, the result being that a heavy, non-porous mass of metallic sulphides is served to the furnaces.   The useful effect of the gangue is a purely mechanical one, and in the same category may be placed the power possessed by some of its varieties of promoting and influencing reaction by mere contact.   The quartz and the acid silicates of the gangue, however, play an important part in the chemistry of the process, since the temperature employed in the lower hearths of the furnace suffices to call into play the power possessed by silica of displacing sulphuric acid from the metallic sulphates present in the charge.

The amount of salt needed for the treatment of the different varieties of ores varies between somewhat wide limits; the quantity, where reverberatory furnaces are used, may be said to lie between 3 per cent. and 7 per cent. of the dry weight of the ore treated.   In practice, in dealing with an ore for the first time, or in case of a change in the ore furnished by a known ore body, the proportion is roughly arrived at by roasting a few pounds of the ore in an experimental furnace; a proportion of salt slightly in excess of that indicated by the experiments is then employed on the large scale, and is gradually reduced under the guidance of the laboratory tests of the roasted product.   In many localities the high price of salt enforces the strictest economy in its use, and in some of such cases it may be advantageous to crush this reagent with the ore, since the perfect admixture produced by this method permits the use of the minimum quantity.   The great objection to this mode of introducing the salt is that the duty of the stamping machinery is thereby lowered, often seriously.   The salt, after leaving the drying apparatus, reabsorbs moisture with great facility, and damps the ore with which it is brought in contact.   For this reason the prevailing practice is to crush the salt separately and to serve the amount needed to the ore in the furnace hopper.

The determination of the loss of silver during the chloridizing roasting of ores offers considerable difficulty.   In laboratory experiments, where elaborate precautions against the mechanical loss of part of the mineral can be introduced, the actual loss by volatilization, under a given set of conditions, can be ascertained; but the value of the results is diminished by the uncertainty as to whether all the conditions of the operation on the large scale are reproduced in these trials in miniature.   In tests on the working scale this loss is approximately estimated in the following manner. A parcel of between 50 tons and 100 tons of ore is crushed to the required degree of fineness, and roasted with the least possible loss of dust. The draught of the furnace in which the experiment is carried out is kept sluggish during the whole period of roasting, and is almost entirely cut off at the moments of running in and advancing the different charges. With the same object in view, the end charges—those nearest the flue—are not rabbled during the experiment. The crushed unroasted ore, the roasted product, and the contents of the dust-chamber attached to the furnace are carefully weighed, sampled, and assayed. The difference between the silver-content of the raw ore served to the furnace and that of the two products obtained from the furnace and its settling-chamber is reckoned as furnace loss. It is expressed in terms per cent. of the silver contained in the unroasted ore. A rough balancing of the account takes place in this mode of evaluating the furnace loss, for, on the one side, no cognizance is taken of the small amount of condensed fume which is recoverable from the colder parts of the flues, and on the other side, the firing and other manipulations of the roasting receive a more than ordinary measure of attention during this experimental work. The routine work of the assay office comprises daily assays of the raw pulp served to the furnaces and of the roasted pulp obtained from them. If the difference of weight between the raw and roasted ore is known from the working tests described, these daily assays give a fair clue to the extent of the furnace loss of silver. For example, if at a given date the yields on assay of the unroasted and roasted pulps are respectively 23 ozs. and 21·8 ozs. of silver per ton, and the loss of weight sustained by the ore during roasting is 4 per cent., the furnace loss of silver will be about 9 per cent.

As a rule, the loss of silver by volatilization is least when the salt is added to the ore at the beginning of the roasting; but in a few cases this loss is reduced when a prolonged oxidizing roasting of the ore precedes the addition of the salt.[1] With many types of argentiferous ore the amount of silver lost by volatilization is much less when the salt needed for chloridizing is added at the start than when it is added at a late stage of the operation of roasting, because the salt undergoes slow decomposition during the whole period of roasting. Part of its chlorine performs useful

---

[1] The time at which the salt should be added to cause the least loss of precious metal by volatilization is discussed by Professor Christy in a Paper on "The Losses in Roasting Gold Ores, and the Volatility of Gold," Transactions of the American Institute of Mining Engineers, vol. xvii., p. 3.

work in chloridizing the metals present in the ore, and part is lost as chloride of sulphur, so that when the stage of the operation is reached at which the temperature is highest, the energy of the reagent as a producer of chlorine is sensibly lowered, and the volume of chlorine to which the finely divided silver compound is exposed is much reduced.

The main influences which tend to increase the loss of silver during the chloridizing roasting of its ores, are, the employment of too high a temperature, the undue prolongation of the roasting operation, and the excessive generation of chlorine, in other words, the addition of the salt in more than the requisite amount, or, at an unsuitable period of the operation. The effect, however, of using much more salt in the working of some varieties of automatic furnaces than is usual in the hand-worked reverberatory furnaces, is minimized by the great shortening of the time of roasting. These exciting causes are only to a partial extent under the control of the operator, a sufficient degree of heat, of duration of the roasting period, and of generation of chlorine, to effect the chloridization of the silver sulphide contained in the ore, must be employed. The effect, however, of such changes of the conditions as are within reach in actual work, may be very great; and is particularly noticeable with respect to changes in the temperature employed in roasting. In the experimental roasting, on the working scale, of three parcels of the same ore, the effect of roasting at temperatures corresponding to cherry red, red and dull red heats, was carefully determined; the furnace losses, beginning with the ore roasted in the greatest heat, were, respectively, 15·9 per cent., 13·7 per cent., and 11·6 per cent. of the silver contained in the unroasted ore. Nearly half the loss occurred in the final chloridizing stage; in a further lot of the same ore, which was discharged before chloridization was complete, the roasting loss was only 6·3 per cent. It cannot be expected that the care bestowed on the exact management of the temperature, in such experimental work as that referred to, will be given to the continuous operation, certain safeguards against the occurrence of over-firing, or its consequences, are therefore provided in the construction of the furnace; the area of the fire-grate is limited as nearly as may be to the working requirements, the arch over the finishing hearth is set high so that the flame does not touch the ore, the fire-bridge is prolonged; where a very delicate ore is worked, it would doubtless be advantageous to further prolong the bridge, and, in fact, to interpose a small chamber between the roasting hearth and the fire-place.

The necessity of carrying out the operation of roasting at the lowest possible temperature is enhanced by the fact that but little of the silver volatilized is recoverable by cheap and readily available means of condensation.   Such part as is saved is found in the deposits and accretions formed on the floors, walls and arches of the colder parts of the flues; these deposits consist mostly of the chlorides and sulphates of the base metals associated with some fine ore dust; assays obtained from them frequently give yields of silver between two and three times greater than those of the ore treated in the furnaces.   The material is deposited in a lightly coherent, flocculent form; it is exceedingly light, so that during the cleaning of the flues it is necessary to partially cut off the draught; in fact, the flues should be cleaned from outside, and the draught entirely cut off during the operation.   The presence of a deposit of this fume on the arches and walls appears to favour the precipitation of fresh coatings.

The use of furnaces in which the stirring of the ore is performed by mechanism, actuated by steam or water-power, has not become general in the north-western States of Mexico; in several of the mining districts the ores obtained are of the highly sulphuretted type, and in many others the absence of good facilities of transport makes the first cost of these motor-furnaces very high.

The use of roasting furnaces, however, in which the manipulation of the ore is effected by machinery, such as the White-Howell, the Bruckner, the Hyde-Bruckner, and the Stetefeldt furnaces, offers great advantages over that of hand-worked furnaces.[1]   The cost of operation is much lower, due mostly to the great saving of manual labour; in remote localities where the supply of labour is intermittent and scarce, the lessening of the cost of this item does not, however, represent the full benefit derived from the use of machinery; the large number of men which would otherwise be needed for the work of roasting is available for other purposes, and this is often a more important matter than the saving of wages.   Great economy of fuel also results from the employment of some of the varieties of the motor-furnaces.   In parts of Sonora, where the hard wood used as fuel costs as much as $8 per cord of 128 cubic feet, it is difficult to over-estimate this advantage.   The rapidity of the action of the motor-furnaces probably also causes a further advantage, in the shape of a diminution of the amount of silver volatilized and lost during roasting.   This loss is, to a certain

---

[1] See also Transactions of the American Institute of Mining Engineers, vol. xiv. p. 576.

extent, a function of the time employed in roasting; if the other conditions which affect this loss, more particularly that of the temperature employed, remain the same, then the shorter the period during which the ore is exposed to the action of heated air and chlorine the less will be the loss; the qualification as to the maintenance of the other conditions is important.

The difficulties that occur with motor-furnaces are, in most cases, due to the presence in the ore of a large amount of sulphur, or of sulphur and arsenic. Since their expulsion is most conveniently carried out by roasting the ore in lump form, the mode of burning in heaps has been experimented with in some few instances. But such additions to the routine of crushing, chloridizing, and leaching do not find favour. The loss of silver sulphate in periods of rain or snow is a great objection to the burning of silver ores in heaps, but the objection does not hold if the operation is carried out in stalls, as these can be easily and cheaply roofed. A considerable proportion of the cost of the extra handling involved in the method is compensated by the lessened cost of crushing, as the burning leaves the ore very friable. The loss of silver during the burning in lump form is, in most cases, small, and the consequent loss in the form of sulphate of silver leached out by rain is easily avoidable; the ore submitted to the action of chloridizing roasting is transferred from the highly sulphuretted, refractory class, to the partially oxidized, or free-roasting class; the operation of chloridizing roasting, that in which the greatest loss of silver is incurred, is therefore rendered much less delicate and more rapid.

The roasted ore is slightly moistened before being trammed to the leaching-vats; it is cooled on a paved area, sprinkled with water from a hose, turned over and mixed with the shovel, and heaped up at the side of the track which passes over the vats. These manipulations are costly, since they require hand-labour, and they are rendered disagreeable and unhealthy by the ore dust which arises during their progress. Improvements in the arrangements of the cooling-floor, with the view of making the work upon it less dependent on hand-labour, and less productive of loss of ore dust, have not kept pace with those effected in the other departments of lixiviation works. On the one hand, the grading of a large space out of the side hill and its roofing are expensive pieces of work; on the other hand, the roasted ore pulp, after being wetted, can neither be raised in an ordinary elevator nor run out of a bin of ordinary inclination; nor, if it contains copper, can it be manipulated in any apparatus constructed of iron. As long, however, as the pulp remains dry, its handling and transport present no difficulty, and it certainly seems desirable that the labour-saving expedients employed in the crushing-mill, such as tipping into bins or vaults, filling into wagons from spouts, moving by horizontal conveyors, or raising in bucket-elevators, should be more freely used in this part of the work. The dropping of the red-hot charges from the roasting-furnaces into a line of vaults, open to the air at one extremity, and communicating with capacious dust-chambers, and the main flues at the other extremity, is merely one step in the direction of utilizing such expedients. A further step would be gained by providing a second line of cooling-vaults, also communicating with the air and with the flues, and filled by wagons which draw from the spouts of the first line of vaults. By such an arrangement the furnaces may be kept clear of the roasted product, and the latter cooled with very little expenditure of labour, and one common cause of the production of dust, the wetting of the roasted pulp before it is quite cold, would be removed. The wetting of ore whilst warm affects injuriously the subsequent extraction of the silver.

The wetting of the cold pulp is most commodiously effected on a narrow strip of ground, bounded on one long side by the railway from the cooling-vaults, and on the other by that leading to and passing over the leaching-vats. The wagons of the latter are filled with the shovel from the ground, but a cheaper mode would be to fill them from a continuous bin by means of the hoe. Moist pulp is slowly destructive of brickwork, so that the bin may be made of wood fastened with trenails of the same material. Certain types of roasted ores are benefited by remaining two or three days in the moist condition. With ores, for instance, containing copper, any silver sulphide which may have escaped decomposition during the roasting is exposed to the joint action of copper chloride, air, and moisture, and is thereby converted into silver chloride. The result of this reaction is an increased extraction of silver by the hyposulphite solution. On the other hand, some ores receive no such benefit, the laboratory tests made on moistened pulp showing no appreciable increase in the percentage of silver soluble in hyposulphites over that contained in the dry pulp; and in such cases the method of charging the vats with dry pulp appears to be worthy of trial.

Leaching and Precipitation.—The roasted pulp, after being placed in the vats, is leached with cold water until the various salts of the base metals, soluble in that menstruum, are removed. The sodium sulphate, resulting from the decomposition of the common salt, and the undecomposed excess of the latter, are also dissolved out. The duration of the operation varies according to the greater or less difficulty offered by the column of ore to the passage of liquids, the amount of soluble salts present, and the quantity of water available. With vats holding 80 tons to 100 tons of ore, it lies between the extremes of eighteen hours and seventy-two hours. The washing is continued until the addition of a few drops of calcium sulphide to a sample of the filtrate produces no precipitate or even turbidity. This first leaching cannot be too thorough, since the salts not removed by it enter into and greatly injure the solution employed in the second leaching. It unfortunately happens that in some of the mines in the littoral and in the foot-hills of the Sierra, the supply of water during parts of the year is inadequate, and either the leaching with water has to stop short of the point of the complete removal of soluble impurities, or the treatment of ore in the dry season has to be curtailed. In such cases the latter alternative should be adopted.

On the first application of the wash-water, care is taken to obviate, as far as possible, the formation of a strong solution of common salt, since silver chloride is soluble in such solution. With this view the water is at first introduced at the bottom of the vat, underneath the filtering medium, and allowed to make its way upward through the mass of roasted ore. As soon as the vat is full, the water is run out at the bottom of the vat and fresh water run in at the top. By this means concentration of the solution in the lower part of the vat is avoided, and such of it as may have begun in the upper part is rapidly diluted. This precaution does not entirely prevent the solving of a small amount of silver chloride in brine, and the first portion of the discharge is therefore run into a special vat and treated with calcium sulphide. When the quantity of copper contained in the ore is sufficient, the base metal water is passed through a series of launders filled with scrap-iron.

The maintenance of the solvent employed in normal condition, as to strength and freedom from impurity, requires frequent and careful attention. The cold dilute solution of calcium hyposulphite enters the leaching-vat through short pieces of hose attached to the main distributing launder. A convenient arrangement is to set the top of this launder flush with the tops of the vats and to pass the rubber tubes through holes bored in the sides of each a few inches from the top; the tubes, 2 inches in diameter, project a short distance into the vats. The solution in the distributing launder is automatically maintained at a constant level, and communication with any of the leaching-vats is cut off by hanging up the projecting extremity of its supply hose.   The operation of leaching is, at first, carried on as rapidly as possible, the discharge orifice below the false bottom being kept fully open.   When the tests show that the amount of silver taken up is much reduced, the volume of solution passed through the ore may be lessened.   With certain varieties of ore, however, time is always saved by circulating a large quantity of solution during the whole period of leaching, even when samples of the solution, after traversing the ore, yield comparatively little silver.   With ores of argentiferous blende, for instance, especially if a little copper is also present, the duration of the leaching operation is considerably shortened by passing through the ore a volume of solution largely in excess of that apparently required.   In order, therefore, that the lixiviation may not be unduly prolonged, on the one hand, or solution unnecessarily circulated on the other, it is advisable, from time to time, to work two or three vatfuls of ore experimentally, and to note the volume of solution circulated, the time employed, and the assay of the spent ore or tailings.

When a considerable percentage of lead, or copper, is contained in an ore treated by this process it is advisable to work with a dilute leaching solution, and to circulate, during the whole period of leaching, as large a volume as practicable of this dilute solution. The use of a solution containing between ten parts and twelve parts of the hyposulphite salt to a thousand parts of water is in every way preferable to that of more concentrated solutions.   The climate in many of the localities in which the process is used is extremely dry, and the evaporation from the surfaces of the receptacles is so great that, unless this matter receives frequent attention, a much higher degree of concentration of the leaching solution takes place than is at all desirable.   These dilute solutions act on the silver salts contained in the ore in preference to the lead salts; the former may be extracted down to the remunerative point before any serious solving of the latter takes place.   The greater part of the silver contained in a vatful of roasted ore is extracted in the first period of the leaching operation.   This is known as the period of sweet solution, because the solution of silver chloride in hyposulphite solutions is sweet to the taste.   The duration of this period is dependent on the richness of the ore in silver, and on the rapidity of the filtration; with a vat holding, say, 80 tons of 25-oz. ore, offering no abnormal resistance to filtration, this period may last between eight hours and twelve hours.   The precipitate yielded during the sweet solution period, even when the ores treated contain a considerable amount of lead or copper, is very rich in silver; dried, and melted in crucibles with a little scrap-iron and borax, it furnishes bars yielding usually about 80 per cent. of silver. On the other hand, the precipitate obtained from base ores, during the remaining period of the leaching operations, is poor in silver; a common yield of the dried precipitate from the final stage of treating impure ores is between 2 per cent. and 3 per cent. of the metal. The hyposulphite solution acts first on the silver salts contained in the roasted pulp, and, when these are almost removed, extends its action to the other salts—notably those of lead and copper—contained in the mass. The greater the degree of concentration of the hyposulphite solution, and the longer the duration of the leaching operation, the greater will be the amount of base metal extracted with the silver. The expediency, when base ores are treated, of employing dilute solutions, and of leaching as rapidly as possible, is therefore apparent. When these two conditions are observed, ores containing considerable percentages of lead and copper are treated with an expenditure on the chemicals of between 60 cents and 80 cents per ton, and furnish a precipitate yielding from 15 per cent. upwards of silver. The spent ore, or tailings, contains usually between 5 per cent. and 8 per cent. of the silver contained in the roasted pulp when ores of less value than 30 ounces per ton are treated; the loss on richer ores is somewhat less. It may be added, that at mines in which the refining of the sulphides obtained from the lixiviation process is attended with difficulty and expense, the silver solution from the sweet period of leaching is precipitated in one set of vats, and that from the finishing period in another set of vats; the precipitate from the former is submitted to the refining operation, and that from the latter is sold as a rich argentiferous lead-copper product.

The silver solution passes to the precipitating-vats through a launder, and has stirred into it a strong solution of calcium-sulphide. During the progress of the stirring, samples of the contents of the vat are frequently tested, and, as long as the addition of a drop of calcium-sulphide causes a tangible precipitate, more of that reagent has to be added. When, on the other side, the addition of a drop of silver solution causes a precipitate, excess of calcium-sulphide is present, and more silver solution has to be run in. As soon as the proper point is reached, the precipitate is allowed to settle and the supernatant liquid is run into the stock-solution of hyposulphite. It is to be particularly noted that the decanting of this liquid as long as any excess of calcium-sulphide is contained in it must be carefully avoided, since the effect of this would be to convert into sulphide the silver chloride already dissolved in the leaching-vats, and to precipitate it in the ore-mass. As a safeguard against such an occurrence the solution is always brought to the safe side of the neutral point; in practice, therefore, the supernatant solution, and consequently the whole stock of working solution, is caused to retain a slight excess of silver solution; it is at the exact point desired when the addition of a drop of calcium-sulphide causes, after a short interval, a thin bluish white film or cloud technically known as the smoke. An excess of the precipitating reagent in the solution is accompanied by unmistakable signs; the liquid becomes milky from the presence of free sulphur, and, if the excess of reagent is at all considerable, the sulphides will not settle but begin to dissolve, and sulphuretted hydrogen is given off. In most mills the operation of leaching and precipitation is attended to by labourers under the supervision of expert foremen. It has often been proposed to carry out the beating and mixing of the liquids in the precipitating-vats with the aid of machinery, but in mills of small or medium capacity this scarcely seems to be needed. During the considerable period needed for filling a vat hardly any attention is given to it, the silver solution and the precipitant are run in simultaneously, and are roughly mixed by turning the stream of the latter into the former and allowing them to fall together; it is only at the end of the operation that stirring by hand and the exact adjustment of the solutions are effected; one man on each shift attends to this and to the leaching work, in mills treating between 40 tons to 50 tons of ore per day.

The contents of the vat are allowed to stand until the precipitate is quite settled, and the supernatant calcium-hyposulphite is then run off to the storage vats. The aperture through which the solution escapes is between $1\frac{1}{2}$ foot and 2 feet above the bottom of the vat, and to prevent the carrying off of particles of precipitate by a too rapid current, the following arrangement is used. A piece of lead tubing is fitted to the orifice, and to it is attached, inside the vat, a piece of rubber hose of length sufficient to reach the top of the vat at about its centre line; two pieces of wood are clamped to the upper extremity of the hose so that it floats in the solution with its mouth slightly immersed. Through this floating discharge the current escapes without causing any appreciable suction on the precipitate lodged in the bottom of the vat; when not in use the hose and its attached float are hung on one side of the vat. The strong silver solution from ore in the earlier stage of leaching yields a heavy precipitate which settles rapidly, whereas that proceeding from nearly spent ore is light and settles very slowly; it is, for this reason, advisable to provide an ample volume of precipitating receptacle so that abundant time may be allowed for settling. Some of the older mills were insufficiently equipped in this respect, and a little fine low grade precipitate was carried off with the solution and arrested at the surface of the ore in the leaching-vats; before shovelling out the tailings a thin layer was raked off and set aside for retreatment.

The great cause of the weakening of the working solution is the oxidizing action of the atmosphere, the contained calcium hyposulphite being converted into gypsum. As long as leaching operations are uninterruptedly carried on, this loss of hyposulphite is compensated—generally more than compensated—by the fresh supplies which are always being formed in the precipitating-vats. But, if active work is suspended, the deterioration of the solution is rapid; a stoppage of work for about ten days greatly impairs its efficacy, and one of a month's duration may render it useless. One of the first requisites, therefore, for the maintenance of the solution in proper condition is continuity of the leaching operations. This requisite is generally upheld without difficulty, since the working of but a small part of the total plant suffices to preserve the solution for a long period; a total cessation of work is rarely called for. The use of a number of vats of moderate size—as against that of one or two very large ones—favours the maintenance of the condition, referred to above, of constancy of regeneration of hyposulphite.

The calcium sulphide employed for precipitating the silver always contains some calcium hyposulphite, which partly accounts for the increase in the volume of the stock of solvent. But the main source of the fresh supplies of calcium hyposulphite which are formed in the precipitating-vats is the double metathesis which takes place between the argentic and other hyposulphites of the silver solution, and the calcium sulphide of the precipitating solution. It follows, therefore, that the higher the degree of concentration of these two solutions, the higher will be that of the regenerated hyposulphite. For this reason it is usual, when ores comparatively free from lead are treated, to make the precipitating solution as strong as practicable; flowers of sulphur and caustic lime are boiled together with such an addition of water that the resulting solution of calcium polysulphide marks 8° to 10°

Baumé. It was formerly customary to use a much weaker solution of calcium sulphide for precipitation, a strength of about 5° Baumé being considered sufficient. With ores rich in silver, and thoroughly well roasted, the consumption of calcium sulphide is sufficiently great to bring about the regeneration of large amounts of calcium hyposulphite, and the stock-solution of the latter is maintained in excellent condition. But, when ores poor in silver and of unfavourable character are treated, the reverse is the case. If, under such conditions, the calcium sulphide is applied in a dilute solution, the fresh supplies of hyposulphite are in too attenuated a form to uphold the strength of the stock-solution. For the purpose, therefore, of being independent of fluctuations in the class of ore submitted to treatment, the practice grew up of employing, under most circumstances, a fairly concentrated solution of the precipitating reagent, the reasons for modifying the practice when considerable amounts of lead, or sometimes of lead and copper, are present in the ore, have already been stated.

The efficacy, as a precipitant, of a freshly prepared solution of calcium sulphide is much greater than that of one which has been some time in stock, since the oxidation of this sulphide into calcium hyposulphite—and finally into calcium sulphate—is always in progress. For this reason it is advisable to limit the size of the apparatus used for the manufacture of the reagent to the approximate requirements of the works, and to make a fresh supply daily. The same necessity of having a freshly prepared article holds good with respect to the burnt lime used in making the calcium sulphide. As the difficulties of transport during the rainy season are great, it is usual to house a large stock of quick-lime before the rains set in, where such a course is practicable; it is, however, better to accumulate a stock of limestone and to burn it, as required, on the spot. A considerable portion of the limestone burnt in the usual manner is not converted into quicklime, but remains as undecomposed carbonate. Since in the manufacture of calcium sulphide caustic lime only is of use, the undecomposed portion of the burnt lime remains as a waste product. This, after being washed, still retains solution of calcium sulphide; it is spread out and exposed to the air until the latter salt is converted into calcium hyposulphite, after which it is heaped under sheds and kept for use. At such times as the stock-solution needs strengthening, a layer of this substance is placed in a vat, and its contained salt dissolved and added to the solution. Whenever it becomes necessary to supplement these methods of maintaining the strength of the stock-solution, sodium hyposulphite is added as required.   A stock of commercial sodium hyposulphite is always kept in store for this purpose: under normal conditions of work its use is seldom necessary ; it is advisable to have this salt packed in metal-lined boxes, as it slowly oxidizes on exposure to air.   A stock-solution of sodium hyposulphite is much more conveniently prepared than one of calcium hyposulphite: operations at new works are therefore begun with the first-named solution, and the same remark applies to the restoration of a stock-solution that may be accidentally lost or destroyed ; the strength used is 12 parts to 15 parts of the crystallized commercial salt to 1,000 parts of water, both by weight, for comparatively pure ores, and 10 parts to 12 parts per 1,000 for base ores ; as the precipitating medium employed is calcium sulphide, a stock-solution of sodium hyposulphite is gradually replaced by one of the corresponding calcium salt.

The lower part of the ore in the leaching-vats retains, after being washed, a large amount of water.   The first portion of hyposulphite solution which is admitted to the wet mass become greatly diluted, and, in order that the whole stock of solvent may not be weakened, this first portion, after traversing the ore, is discarded.   The lixiviation of a vatful of washed ore is usually begun by running hyposulphite into the vat until it is about one-third full, the discharge orifice is then opened and the diluted solvent is run to waste, the feeding supply being kept on.   As soon as the discharge, on being tested with calcium sulphide, shows traces of precipitate, it is run into a special receptacle. When the samples from the discharge yield tangible precipitate, the solving of silver chloride has begun, and the liquid is run into the precipitating-vats proper.   The liquid in the special receptacle, after receiving an addition of calcium sulphide for the recovery of the small amount of silver it contains, is run to waste.   At the end of the lixiviation, the ore-mass is saturated with hyposulphite solution, and the reverse of the above procedure is carried out ; water is admitted to the spent ore, takes up the hyposulphite, and is run into the stock of solution.   The plan described of running to waste a few cubic feet of the hyposulphite solution each time a fresh batch of ore is treated, besides serving to prevent the dilution of the whole stock of solution, serves also to keep within bounds the concentration of foreign salts within it.   The presence, even in coarsely crushed ore, of a considerable proportion of slime, renders the extraction of the intermixed common salt and sodium sulphate, by leaching with cold water, slow and tedius ; these salts are not entirely removed in this first leaching, and traces of them are therefore taken up by the hyposulphite solution used in the second leaching. Further, sodium chloride or calcium chloride, as the case may be, must of necessity be formed in the stock-solution as a consequence of the reaction which attends the solving of the silver chloride in it, or of the solving of basic chlorides of other metals; this reaction may, in the case of the sodium-salt, be thus expressed, $2NaS_2O_3 + 2AgCl = Ag_2S_2O_3 + 2NaCl$. The entry of foreign salts into the stock-solution would gradually render it useless as a solvent.

In the Russell process the ores are first treated with a solution of sodium hyposulphite, and then with a solution of a double hyposulphite of sodium and copper. The latter solution is made by adding sulphate of copper to the former. It exercises an energetic solving action on metallic silver and on various salts of silver which are not acted on, or only slightly acted on, by an ordinary hyposulphite solution. Its action on silver sulphide, whether existing alone or associated with antimonial and arsenical sulphides, is particularly important. When the chloridizing roasting of an ore is imperfectly carried out, these compounds, in an undecomposed or only partially decomposed state, are of frequent occurrence. Under the treatment of the ordinary lixiviation process they resist the action of the solvent employed, and are wasted in the tailings; whereas, under the treatment of the Russell process, they are to a large extent decomposed and brought into solution. With ores which have been submitted to a thorough chloridizing roasting, the use of this process offers no advantages over that of the ordinary process. In fact, the ordinary hyposulphite solution is a better solvent of silver chloride than is the double hyposulphite solution. Also, when cupriferous ores are treated, the advantage of the Russell process over the Patera process disappears. The cupric oxide and cupreous chloride, formed during the roasting of the ore, are not removed by the leaching with water, but are dissolved in the calcium hyposulphite solution of the second leaching, and confer on it the properties of double hyposulphite solution. If, therefore, copper is present in the ores treated, the solvent employed in the Patera process is, to all intents and purposes, a solution of cupreous-calcium-hyposulphite. The fact that the Patera process is able to treat all copper-bearing argentiferous ores with most excellent results was known from the earliest days of its use; but the reason why the presence of salts of copper influenced the extraction of the silver was not known until the results of Mr. Russell's exhaustive experiments on the solubility of combinations of silver, in solutions of double salts of sodium and copper hyposulphite, were published [1] by Mr. Stetefeldt. The fact that with many types of ore the successful working of the Russell process is less dependent on a perfect chloridizing roasting than is the case with the Patera process, gives the former a great economic advantage over the latter. At the Yedras mines this advantage was very great—so great as to cause the abandonment of the use of the older process and the adoption of the newer one. The saving effected by the change amounted to $6 or $7 per ton, less the charge for royalty. A detailed account of the comparative working of the two processes has been given [2] by Mr. Rockwell. The nomad character of the labour at these mines rendered thorough chloridizing roasting impracticable, and the ore dealt with did not contain a vestige of copper. On the other hand, at some of the pioneer lixiviation works in Sonora, where these two important conditions were reversed—the chloridizing being carried out with trained roasters, and the ores treated being cupriferous—the experiments made with the Russell process showed the same results as those obtained from the Patera process. By the introduction of the Russell process the sphere of usefulness of lixiviation methods of treating silver ores is considerably enlarged.

The use of the extra solution is the distinguishing feature of the Russell process. But, to the operator by the Patera process, the fact that the calcium hyposulphite and calcium sulphide used in that process as solvent and precipitant are, in the Russell process, replaced by sodium hyposulphite and sodium sulphide, is of more immediate interest. Mr. Stetefeldt, in his essays on the Russell process, points out that the method used therein for precipitating as carbonate the lead contained in the silver lixivium prohibits the use of calcium hyposulphite as solvent, but that process loses nothing by this inhibition; that, on the contrary, the use of these salts of calcium is, in every way, less advantageous than that of the corresponding salts of sodium. Sodium hyposulphite as solvent, and, consequently, sodium sulphide as precipitant, were used originally with the Patera process; they are now used with the Patera process at some of the mines of the Barrier Range, New South Wales. But, at the time of the first working of the process in Sonora, the means of transport to manufacturing centres were very costly, and it was the object of endeavour of those engaged in

---

[1] "Transactions of the American Institute of Mining Engineers," vol. xiii. p. 47.

[2] "The Engineering and Mining Journal of New York," vol. xlv. p. 86.

mining and milling to carry on their operations as far as possible with such supplies as could be obtained in the country itself. As long as sodium sulphide was used as precipitant, the caustic soda needed for its manufacture had to be brought in from abroad, whereas the caustic lime needed in making calcium sulphide was obtainable, in the majority of cases, in the vicinity of the works. When it is considered that beyond the high cost of transport opportunities of effecting it were both infrequent and unreliable, and, further, that caustic soda is an inconvenient substance to convey on pack-animals, it will be seen that strong inducements existed for the preferential use of the calcium salt. The custom of using calcium hyposulphite and calcium sulphide, therefore, became established, and, coincidentally, came in the belief that, of the hyposulphites of the alkalies or of the alkaline earths, those of lime were best suited for use in this branch of metallurgy. It is, however, true that a solution of calcium hyposulphite is not superior to a solution of sodium hyposulphite as a solvent, whilst it is inferior to the latter in stability. And so with the corresponding precipitating solutions, calcium sulphide possesses no point of superiority over sodium sulphide, and, in at least two important respects, consumption of material when in course of manufacture, and stability when in use, is inferior to it.

Treatment of the Sulphides.—The precipitate which is obtained by the lixiviation process is, after being thoroughly dried, either melted and run into bars of impure bullion, or subjected to a mixed scorification and cupellation process. In some few cases, however, it is simply packed in bags and boxes, and shipped for sale to foreign refineries.

The drying is in nearly all cases effected in reverberatory furnaces. These are made quite narrow, a width of about 7 feet being commonly adopted. The length given to them is dependent on the amount of material to be dealt with; frequently it is about 15 feet. Since the object aimed at is to expose the precipitate, not to contact with flame, but to the action of a current of hot air, the fireplace is made very small. A chamber is interposed between the drying-hearth of the furnace and the fire-bridge, and the products of combustion of the wood burnt on the grate are mixed in this chamber with heated air drawn through channels in the brickwork. The sole of the hearth, which receives the wet precipitate, is placed over a vault which forms part of the dust-chambers; it is dried and slightly heated by the current passing through this vault. In view of the richness in silver of the precipitate, the use of muffle furnaces for this drying operation has been suggested; but the furnace described above is cheaper to build and to keep in repair, and does its work satisfactorily. It might, however, be an improvement to cause the products of combustion to pass under the drying-hearth before entering it. When the precipitate is thoroughly dry, the temperature is sufficiently raised to ignite the free sulphur which is always present. As soon as this point is reached, all the fuel is raked out of the fireplace, and the combustion proceeds unaided. The mass is turned over once or twice, care being taken to almost arrest the draught of the furnace during this operation. The custom at one time prevailed of not only burning off the free sulphur, but of driving off part of the combined sulphur by calcining the product. This practice has very largely, and very properly, been abandoned; the substance is far too rich in silver to render such treatment judicious. The oxidation is far better effected in the cupellation furnace, where the liberated silver is at once seized upon and protected by metallic lead. On the other hand, it may be noted that precipitate is frequently dried by being pressed in filter presses and exposed to a current of air heated by contact with pipes through which steam is passed. If precipitate dried in this way is to be treated by cupellation, it seems desirable to first burn off the uncombined sulphur.

The process of refining the precipitate was, at Promontorios, carried out in cupellation furnaces of the German pattern, the hood being movable and the hearth fixed. The latter was circular in plan and $6\frac{1}{2}$ feet in diameter. Wood was employed as fuel, and the marl used for the bottom was obtained by crushing a mixture of broken limestone and clay through a sieve with sixteen holes to the lineal inch. After the bath of lead, which weighed about 5 tons, was freed from the first impure litharge, the blast was kept on until the surface was again completely covered with litharge. The blast was then cut off and the draught greatly reduced, and between 150 lbs. and 200 lbs. of precipitate introduced with a ladle. Clean litharge, to the extent of three times the weight of precipitate taken, was then spread over the latter. As soon as the added material reached the pasty condition, the draught was restored, the blast applied, and the fire slightly urged. When the scum, which resulted from the fusion of a charge, was nearly liquid it was skimmed off, litharge was again allowed to form over the bath of lead, and a fresh quantity of precipitate was charged as before. In order that the scorification method of treating silver precipitate may work smoothly, and with a minimum of loss, it is essential to carry it out in furnaces of ample dimensions. When this requirement is observed, the mixed sulphides may be brought in contact with many times their weight of litharge. The layer exposed to the oxidizing action of the blast is thin and penetrable, the decomposition of the sulphides is therefore rapid and complete. Part of the copper contained in the precipitate is absorbed by the bath of lead, but in large furnaces is so diluted as to have comparatively little ill effect. When the yield of copper in the precipitate is large, the lead bath is maintained in good condition by, from time to time, interrupting the charging of precipitate, and by feeding soft lead into the bath until its surface becomes bright. With the view of keeping at the lowest point the amount of copper which is taken up by the lead, the refining operation is conducted at the lowest temperature practicable. The precipitate, when mixed with large amounts of litharge, is fairly fusible, but it is not necessary to employ such a temperature as to completely liquefy the slag or scum which results from its fusion, or partial fusion. The commercial product obtained from the above described scorification is rich argentiferous lead containing some copper, and its subsequent treatment is simply one of cupellation. This is effected in a smaller furnace of the same pattern, or in the English test-furnace. The litharge and the scum which are produced in the refining-process are reduced in blast furnaces. At Promontorios and at Minas Nuevas the lixiviation process and the lead-smelting process were both in use, and the lead products from the refinery were added to the smelting mixture; their contained copper was collected in the mattes. The dried precipitate submitted to the refining operation at the Promontorios works contained about 20 per cent. each of silver, copper, and lead. These proportions, of course, varied over wide limits. The yield of silver lay between the extremes of 14 per cent. and 35 per cent., and the yield of copper between 15 per cent. and 27 per cent. It may be noted that the average fineness in silver of the precipitate, obtained by the lixiviation process from ores which are comparatively free from lead and copper, is from 45 per cent. to 55 per cent.

The lixiviation process is not applicable to the treatment of gold ores. Of the small yield of this metal contained in the silver ores submitted to the action of the process, from 50 per cent. to 70 per cent. is obtained.

One of the useful applications of the Patera process is the treatment of the argentiferous zinc blende, which is obtained by the hand-sorting and mechanical sorting of mixed ores of blende and galena. The important point in connection with the treatment of these sorted products is the degree to which it is necessary to carry the separation of the galena. With the ores treated at the Promontorios works, the separation of the galena from the blende by the hand-pickings, the sizings and the jiggings which were resorted to, was not perfect; blende was retained with the galena separated for smelting, and galena accompanied the blende set aside for treatment by lixiviation; the slimes, in particular, contained a considerable proportion of galena. It was found that, what may be called the permissible content of lead, in the case of these ores, for economical treatment by the Patera process, was passed when 8 per cent. of that metal was present in the roasted pulp: with this amount of lead the consumption of precipitating reagent was equivalent to 11 lbs. of sulphur for each ton of ore treated. The average consumption of sulphur at these works was 6·7 lbs. per ton of ore treated, so that the average content of lead in the roasted pulp was well below the point named above. A comparatively small proportion of the ores extracted from the mines consisted of a fine-grained variety of mixed ores of blende and galena; the two sulphides could not be separated by jigging, and these ores, in admixture with large amounts of ores poor in zinc, were therefore smelted. The Patera precess is not competent to deal with those varieties of zinc-lead-sulphide ores in which these sulphides cannot be separated—or, rather, partially separated—by mechanical dressing. An exception to this statement occurs when the yield of lead in this practically unsortable variety is low; the process is competent to deal with these mixed ores when this condition is fulfilled. Its use for this purpose has largely increased of late; the Author found it in use, in the treatment of argentiferous blende ores carrying some galena, in the Peruvian Andes some three years ago, and was informed that it had also been introduced into Chili for a similar purpose; its somewhat extended employment in New South Wales has been, for the most part, in connection with the treatment of this same class of ore. When its use is practised with the segregated product from the dressing of these mixed ores, the mode of preparing the slimes from the dressing works, for entry into the chloridizing furnaces, calls for a little attention. At Promontorios, the slimes produced during the crushing of the mixed blende and galena were not dressed; long-continued experiments showed that a series of washings on endless belts and inclined planes, failed to remove the silver sulphide contained in them; they were merely roughly sorted by being passed over strips, the headings obtained were added to the smelting ores, and the whole of the residue was settled in reservoirs and added to the lixiviation ores. These slimes formed in drying hard lumps, which were impermeable alike to the chloridizing action of the roasting process, and to the attack of solutions; they were prepared for entry into the chloridizing furnaces in the following manner. The contents of the slime-pits were run on an inclined paved surface and allowed to drain and partially dry, and were then fed into a Pacific dryer. This apparatus differs but slightly from the rotating-cylinder roasting-furnaces of the continuous action type; its cylinder is made of cast-iron and is not protected by a brick lining. The dimensions of the cylinder are: length 18 feet, diameter at the feed end 3 feet and at the discharge end 4 feet; it is set with its axis horizontal and the inclination needed for the advance of the ore is furnished by its taper; for such purposes as the drying of slimes the dimensions above given might advantageously be increased. The dried product passed into a Bruckner ball pulverizer—a cylinder in which the material treated is rotated together with a number of iron balls, and to part of the periphery of which a sieve is attached.

Both these appliances worked economically; the dryer is, for convenience of carriage, made in short rings, but, when the mode of transport is by pack-mules, it is further "sectionalized"; at points where freight is extremely dear a convenient substitute for this apparatus is a long reverberatory furnace, the hearth of which has a very steep inclination from the flue end to the point of discharge.

Cost of Working.

The cost of ore-reduction by the lixiviation process is influenced by and, in great measure, dependent upon the chemical and physical constitution of the ore, the local conditions, such as the prevailing rate of wages and the price of supplies, and the facilities for handling the ores and products offered by the plant. The most important item of cost is that of the chloridizing roasting of the ore, and this fluctuates between very wide limits. The influence exercised by the character of the ore is most marked on this branch of the expenditure, since, when ores of a favourable type are roasted, the daily capacity of a chloridizing furnace may be nearly twice that which can be reached when ores of the opposite type are roasted. The influence exercised on the cost of roasting by variations in the local conditions is less marked, still it is considerable. The most important of these variations is that in the quality of the labour obtainable in the different localities; in some parts the furnace work is carried on with trained and reliable roasters, and in others with untrained and unreliable labourers; the rate of wages paid for this work does not vary much, commonly it is between 75 cents and $1 per day. The only articles of supply, the price of which affects to any large extent the total cost of roasting, are firewood and common salt. The price of the former may vary, at different points, between $3 and $8 (Mexican silver dollars) per cord of 128 cubic feet, and that of the latter between $15 and $40 per ton of 2,000 lbs.; the principal factor which governs the price of each is the cost of freight from the sources of supply to the reduction works; the consumption of wood is between 0·15 cord and 0·20 cord per ton of ore roasted. In the leaching department, the main item of cost is that of sulphur, the consumption of which varies, according to the nature of the ores treated, between 4 lbs. and 9 lbs. per ton of ore; its price is between 5 cents and 8 cents per lb. The amount of quicklime used is between one-and-a-half and two-and-a-half times that of the sulphur, and its price is generally about 1 cent per lb.; sodium-hyposulphite is rarely used. Under the varying circumstances enumerated, the total cost of treating ores ranges between $7 and $13 per ton. As is to be expected, the aggregate of circumstances which moulds the cost of the work is rarely entirely favourable or entirely unfavourable, one set of conditions roughly balances another set, so that the average cost of treatment, at works treating between 40 tons and 50 tons per day, is about $8 to $9 per ton in the coast districts, and $11 to $12 per ton in the mountain districts. The process is a favourite one with owners of small mines; in the reduction works with which such mines are equipped, say with a capacity of 20 tons per day, the ratio of cost of management to that of total cost is a high one; such works are, however, usually directed by the owners. The most important economy is that to be expected from the introduction of improved methods for roasting the ore; the adoption of mechanical roasting furnaces, or the use of reverberatory furnaces of large dimensions, would undoubtedly effect a considerable saving in the cost of working.

The Paper is accompanied by three drawings, from which Plate 2 has been prepared.

(*Paper No. 2912.*)

"Mining and Treatment of Copper Ore at Tharsis, Spain."[1]

By CHARLES FREDERICK COURTNEY, M. Inst. C.E.

IN the development of the Tharsis mines four periods of well-defined activity can be traced, the Prehistoric, the Phœnician, the Roman and the Modern.

During the first, the emergence from the stone period is indicated by the quantities of stone implements, made from the hard porphyry of the surrounding country, which have been from time to time found in the mines and caves of the district.

The second period occurred during the Phœnician occupation of Spain, which, according to Josephus, was 240 years before the building of Solomon's Temple, or about 1,200 years B.C. Gades (Cadiz) was then one of the most important ports, and only six hours' sail from Huelva; it may be assumed that the Phœnicians were at or near the Andalucian group of mines about this time. The period of their domination may be also ascertained, for the weakening effect of the inroads made by the Assyrian monarchs was so disastrous to their trade that their great power in their colonies was broken. The Carthaginians, taking advantage of their difficulties, are found to be making inroads into Spain, and depriving the Phœnicians of their authority, about 650 B.C. There is no distinct evidence of the mines having been worked about this time, although the fact of several heaps of a rough and spongy class of scoriæ being in close proximity to the lodes, and always near an abundant spring of water, has strengthened the belief that these heaps are the refuse of Phœnician smelting. The complete absence of Carthaginian mining enterprise in the province indicates that the Romans, in becoming masters of Spain, must have rediscovered the mines; and that between the Phœnician and the Roman domination, 200 years B.C., or a period of 400 years, no mining work was carried on in this district.

During the third, or Roman period, the population drawn together for the purpose of trade and mining numbered millions, and remains can now be seen of an industry that was undoubtedly

---

[1] The discussion upon this communication was taken in conjunction with the preceding and the following Papers.

colossal; roads, amphitheatres and aqueducts abound, as well as numberless articles of jewellery, copper pots and pans, coins, mining-lamps, and water-wheels. It is difficult to imagine how so great an amount of ore as that represented by the slag-heaps which cover the ground, was extracted through the tortuous galleries at present existing. Many miles of railway and road have been ballasted with this slag, and still the volume is not perceptibly reduced. These slags are as fresh and clean in outline now as they were when run from the furnace 2,000 years ago. It is indeed strange that so vast an industry as was here established, of 500 years' duration, could so completely have ceased. For 1,000 years there was again an absence of work in the district, partially accounted for by the Moorish domination of 410 A.D.

The renewal of activity by the Spaniards marks the opening of the fourth or modern period, and may be traced to the desire for riches induced by the discovery of America. In the year 1866 the present company was formed with a capital of £300,000 to take over the mines and property from a French company; and, as the working of the lodes developed, it was found necessary from time to time to increase the capital, which now stands at £1,250,000. Since the commencement of operations by the present company 508 per cent. has been paid in dividends, and £1,884,885 written off the property plant, &c.

The composition of Roman slag is compared with that of modern slag in the following Table. Phœnician slag differs from the Roman and modern slags in containing between 2 per cent. and 3 per cent. of copper.

—	Roman Slag.	Modern Slag.
	Per Cent.	Per Cent.
FeO	50·81	46·59
Fe	0·46	2·40
Fe₂O₃	4·48	Trace
Al₂O₃	6·85	6·80
CuO	2·23	0·90
Cu	0·12	0·47
Zn	None	0·58
PbO	1·00	None
S	None	1·60
KNaO	0·28	1·06
MgO	0·31	0·38
Silica	32·80	38·15
Moisture	0·16	None
Combined moisture	0·89	None
	100·39	98·93

GEOLOGY OF THE DISTRICT.

The property of the Tharsis Company consists of the Calañas lode, from which the export ore is at present drawn, the six lodes in Tharsis, the main establishment, and the newly acquired Lagunazo mine about 5 miles distant. The Calañas mine, about 18 miles distant, is connected with the main establishment by railway, and a line, 29 miles in length, connects Tharsis with the port at Huelva.

To the south of the Sierra Morena, and forming part of that range, there is, extending from near Seville and running into Portugal, a zone of clay slate between 110 miles and 120 miles in length, which encloses the enormous deposits of cupreous iron pyrites at Rio Tinto, Tharsis, and San Domingos. The course is in a north-westerly direction; and parallel to the deposits of ore, dykes of quartz-porphyry occur, there being in some parts a difficulty to distinguish the slate from the porphyry. These dykes and intrusions are, however, generally prominent, and the enclosing rock is believed to be of either Silurian or Devonian age. The lodes are about 800 feet above the level of the sea, and the colouring of the surrounding country and their immense iron capping form prominent features of the district. The masses of ore are of a more or less lenticular form, parallel to the enclosing slate, and of great length and width, being as much as 2,600 feet long, 65 feet to 500 feet wide, and in most cases of great depth. The upper portion, having absorbed the copper liberated by the oxidation of the outcrop, contains as much as 3 per cent. or 4 per cent. of copper, whilst at a depth of 300 feet to 400 feet, unoxidized specks of yellow ore are found, giving the mass only ½ per cent. of copper. In this form, the ore, both from poverty in copper and compactness of structure, becomes useless for either export or local treatment. Owing to the rapid decomposition of the pyrites, the existing ironstone outcrop or capping marks very perfectly the underlying mass; and the enclosing slate, by the action of the acid salts, has become soft and of a yellow, red, and black colour. When the hydrated oxide of iron of the capping is associated with silver and other metals, these may be considered as an inset to the occurrence of cupriferous pyrites at greater depth. *Fig. 1*, p. 44, shows the form assumed, and the connection between the gossan and the lode. One of the lodes of the Tharsis group, however, merits special mention, as it is, at least in this district, unique, consisting of slate that has been impregnated by cupreous liquors. Such deposits are restricted to sites that have received the natural liquors produced by the oxidation of pyritic lodes at their outcrop, and have originated in the action of these cupreous liquors upon slate containing disseminated sulphide of iron and zinc. The most characteristic indications are depositions of ferric oxide, either in blocks forming a conglomerate, or as films upon slate or other rocks.

The whole mass, of 2,300 feet length, with an average width of 197 feet, and depth of 98 feet, contains very little copper, its amount seldom exceeding 1½ per cent., and corresponds exactly to the outer case of the pyritic lodes, where the ore, being in contact with the slate, has caused the latter by impregnations to become soft. Such deposits are very easily and economically worked, and permit of profitable extraction with a copper contents of a little

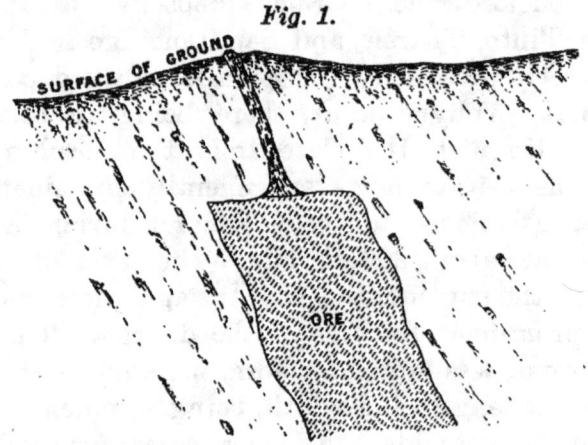

Fig. 1.

over ½ per cent. The composition of the ores is as shown in the following Table.

| | Cupriferous Iron Pyrites. | | Cupriferous Schist. |
	Export Ores.	Local Treatment Ores.	Local Treatment Ores.
	Per Cent.	Per Cent.	Per Cent.
Iron	43·10	42·80	0·10
Sulphur	49·50	48·30	0·15
Copper	2·50–3·00	1·50–2·00	1·00–1·25
Lead	0·30	1·30	..
Zinc	0·80	1·80	0·05
Arsenic	0·40	0·70	..
Water	0·50	0·80	3·15
Silica	0·60	0·70	93·60
Oxygen and various metals . .	1·20	1·10	..
	99·40	99·50	98·30
Specific gravity	4·50	4·50	2·25

METHODS OF MINING.

Three methods of mining have been employed at Tharsis for the extraction of the ore. The pillar-and-stall system has been largely adopted, and by it somewhat over one-third of the entire mass can be extracted without the necessity of support or danger in working, the ore being very compact. Each floor has a height of 33 feet, and parallel galleries are driven through the entire length of the lode, 16 feet by 16 feet and 33 feet from centre to centre. A partition is left 16 feet thick, which is afterwards broken through in stalls of 16 feet width and 33 feet centres; the stall of one partition corresponding with, or facing, a pillar in the opposite partition. Great care has to be taken in the surveying of the various levels, as the safety of the lode depends upon the pillars falling vertically over or under each other on their superior or inferior floors: large faces of ore are exposed, affording freedom for the men in winning, and enabling the extraction to be continued with regularity. Each floor has a fall of not less than 1 in 150 towards the extraction shafts, which is found sufficient for both drainage and easy traction. The ore when broken down by the miners is loaded by a separate gang into small wagons, of 1 ton capacity, running in a 2-foot 6-inch gauge line, and made up into trains of three or four wagons for mule traction to the shafts. Each miner wins 2 tons per day, and a filler averages 20 tons per day of eight and a half hours, the consumption of dynamite No. 1 being about 0·175 lb. per ton. The miners and fillers are generally on contract and earn 3s. and 2s. 6d. per day respectively. By this system the ventilation is excellent, there are therefore no delays from dynamite fumes after blasting. The dryness of the ground has always facilitated underground working, as not more than 30 gallons per minute of liquor per day is given by these large lodes.

The large amount of ore that was left in the lodes after working on the foregoing system led to the contemplation of other methods; and finally, after much thought and consideration, a bolder scheme was inaugurated—that of removing the overlying rock. The open casting of these lodes has been at work for many years, and some of the largest excavations in the world may now be seen in this province. The dip of the ore, not being more than 60°, assisted the open-cut system of working; whilst the more economical mining of the ores afterwards has permitted a very large amount of overlying rock to be removed profitably per ton of ore rendered available. Six hundred and fifty-four thousand cubic yards have been removed from the surface of one lode within a year. The method employed, *Fig.* 2, is that of dividing the over-burden into floors, 33 feet in height, leaving a slope of 60° and a bank of 13 feet width; this gives an average batter of about 45°, allowance being made for the varying conditions of the hanging- or foot-walls. The hanging-wall permits of a batter of $3\frac{1}{2}$ in 10 being given as against $5\frac{1}{2}$ in 10 for the foot-wall. There is, however, no set rule, the slopes in all cases being governed by the strength or compactness of the enclosing rock; but this condition and the copper contents are conjointly the factors which fix the cube that can be allowed per ton of ore made available. From the large open cuttings, many banks being over 1,600 feet in length, a miner will blast and break down 250 cubic feet per day of eight

Fig. 2.

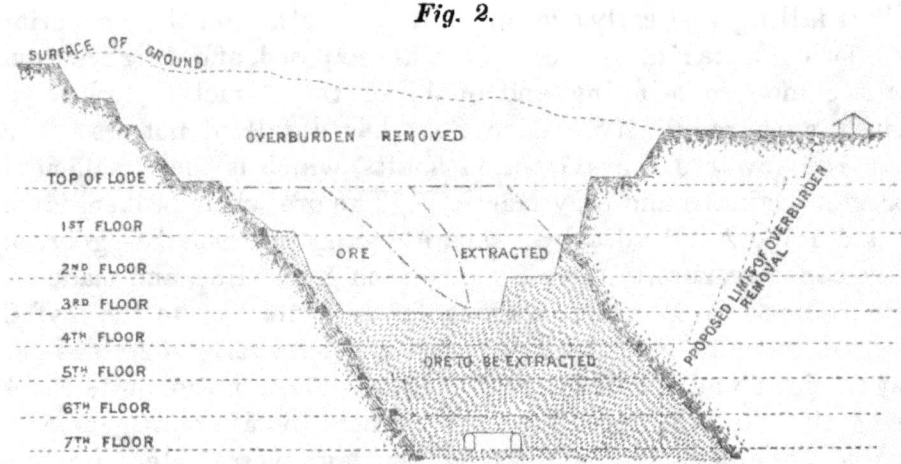

SECTION ACROSS A MAIN LODE SHOWING ORE AND OVER-BURDEN REMOVED.

Scale, 240 feet to 1 inch.

and a half hours at a wage of 2*s.* 6*d.*, whilst a filler at 2*s.* $2\frac{1}{2}d.$ per day, assisted by a carrier to wagons at 1*s.* $3\frac{3}{4}d.$, will give 424 cubic feet during the same time. As in this case the work is generally by task rather than by contract, a closer supervision is required over the consumption of dynamite, the average amount used being 0·168 lb. No. 3 per cubic yard. An excavation that advances with the line of stratification will require one half more dynamite than if advancing at right angles to it; it is therefore always advisable in open casting a lode of considerable length and ample width to drive a trench in advance, giving in this way a minimum of work with the stratification and a maximum at right angles to it. Whether the dip be great or slight, the work should always, where practicable, be commenced on the hanging side, working towards the foot-wall, there being a considerable difference in the consumption of dynamite in favour of working with, as working against the dip. After the removal of the over-burden the extraction of the mineral becomes much easier and consequently less costly. Each man will blast down 8 tons at a wage of about 2*s*. 8*d*. per day, as against 2 tons per man in close workings at a wage of 3*s*. per day, loading in both cases being about the same, whilst the consumption of dynamite No. 1 is about 0·099 lb. per ton in the open cutting, as against 0·175 lb. in underground workings. The advantage is, therefore, that the whole of the ore uncovered by the removal of its over-burden is rendered available for extraction, the lower cost of extraction more than paying for the removal of the superincumbent rock.

The third method of mining is more or less an adaptation of the long-wall system, and permits of the whole of the ore being

Fig. 3.

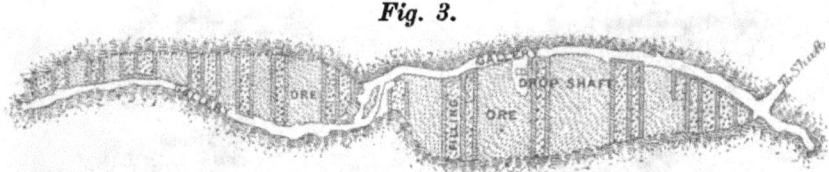

PLAN SHOWING METHOD OF EXTRACTION BY BUILDING AND FILLING.
Scale, 340 feet to 1 inch.

removed by underground working. The lode is divided longitudinally at equal distances by diaphragms or walls, starting at the bottom level and working upwards, between each wall a space being left equal to their thickness, afterwards to be filled with loose rock or earth as the overhead extraction continues. This system, though expensive in working, has been adopted for the extraction of the ore from one of the lodes the form of which lent itself readily to the operation. The usual underground working would have given, as before stated, about one-third of the whole mass, the richness of the ore warranting a more expensive system of working though not the removal of its over-burden. This mass of ore, known as the Poca Pringue lode, contains about 3·75 per cent. of copper, its length being 679 feet, and having an average width of 52 feet, being evidently a spur from the adjoining lode, although separated therefrom, as it runs out at a depth of 112 feet. The lode is divided into floors 33 feet in height, and a gallery at each floor runs along one side of the mass, assisting ventilation and enabling the extraction to be carried on easily. Operations begin at the lowest level or floor, the entire length being sub-divided into spaces of 4 feet alternating with those of 8 feet in width, the 4-foot spaces being reserved for building in dry stone and the 8-foot spaces for filling. Galleries are driven across the lode at the bottom of 16-foot width, and the excavation continues upwards by overhead working. As the ore is removed the dry-stone walls are built along each side of the excavation, the centre portion being filled with waste. When the ore appertaining to this section is extracted there is left in its place a diaphragm from the bottom to the top of the mass composed of two dry-stone walls enclosing rough filling. Whilst this section is being worked, other

Figs. 4.

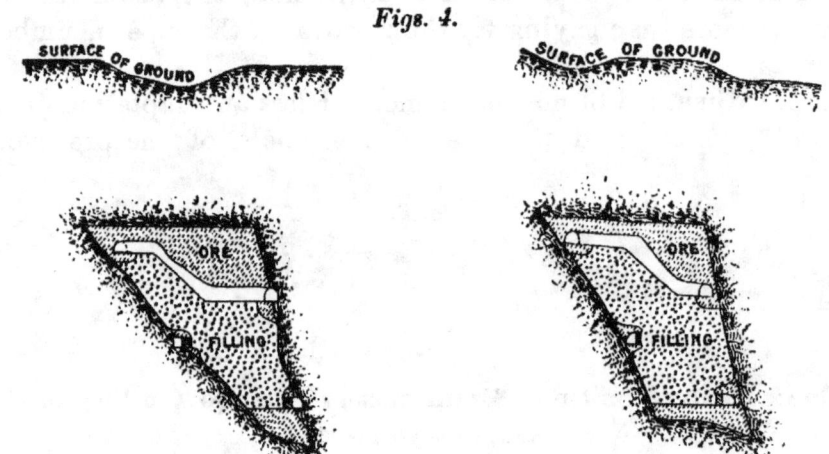

SECTION SHOWING METHOD OF EXTRACTION BY BUILDING AND FILLING.

Scale, 240 feet to 1 inch.

portions of the lode are being attacked in the same manner, the upper exterior galleries being used for the supply of stone, &c., for building and filling, whilst the lower galleries serve for the loading of the extracted ore. The stone for building and the waste for filling are obtained from the open cutting of the adjoining lode, being selected during removal and deposited near an automatic drop-shaft by which it is passed when required to the workings below. No difficulties are experienced from any movement or creep of the hanging-wall. The whole operation of mining, dry-stone packing and filling, with the present low market prices for copper, sulphur, &c., requires about 1·50 per cent. to 1·60 per cent. of copper per ton of ore obtained to cover the costs of working.

MACHINERY.

The shafts vary in size considerably, some being large enough for a double cage, and others for a single cage only; the output, however, for each shaft may be considered as between 900 tons and 1,000 tons per day, drawn from an average depth of 295 feet. The shafts are fitted with ordinary cages, safety-hooks and automatic gong indicator in the engine-room for the starting and delivery of the wagons. At the shaft mouths there are extensive platforms, along one side of which a series of tumblers is arranged by which the wagons are turned completely over by their own weight, righting themselves again when empty. To the tumblers are connected shoots at an angle of 36° in the form of riddles of ¾-inch spaces, to allow the mineral for export to be cleansed of its smalls, which amount to about 17 per cent. of the whole weight. The smalls are sent to the local treatment ground for the extraction of their copper, and the reduction of the liquors before precipitation; other shoots are fitted with the Blake type of crushers or stone-breakers by which the ore is reduced to the required size for local treatment. These crushers have jaws 24 inches wide, and are capable of crushing 40 tons per hour at a cost of $2\frac{1}{4}d$. per ton. The channelled and ribbed jaws of chilled iron are reversible and withstand the crushing of 1,700 tons each; they have therefore to be constantly replaced and provision is made for their easy removal. An automatic balance-feeder is also attached to the end of the riddle where the ore drops into the crusher, to prevent the whole wagon-load rushing into the jaws and causing them to jam. By these arrangements very little labour is required to keep the crushers constantly fed, the ore after being crushed drops through a funnel into the wagons waiting for its reception beneath, and is ready for weighing before being drawn off in trains of twenty wagons to the local treatment ground.

Each shaft is fitted with a 12-inch bucket-pump for dealing with the drainage water, although the ground, as before stated, does not yield much water; the extent of the open cutting accumulates large quantities, which run in upon the lower floors suddenly, and sometimes requires heavy pumping for a few hours, especially during wet seasons or thunderstorms. As these liquors are very acid, and give great trouble by rapidly eating into the metal of the pumps, an alloy containing one part of antimony, two of lead, three of tin, and fifteen of copper for cylinder buckets, valves, etc., though difficult to cast, has been found to give the most excellent results, whilst for bolts, nuts, etc., one part of lead, two of tin, and seventeen of copper, is preferable to withstand blows or strains, being more ductile.

The workshops are well fitted with modern machinery, and all the smiths, mechanics, fitters, boiler-makers and carpenters are Spaniards. All locomotive and wagon repairs are carried out at the mines.

Treatment of the Ore.

The old and simple method of extracting the copper was by calcination, the sulphide of copper during the process being turned to cupric sulphate, and the ore to a spongy condition by the consumption of its sulphur; the copper was then very readily washed out with fresh water. A large area of ground was occupied by the ore both in active calcination and that which had been calcined, being first conveyed to the site, either crushed by the crushers at the pit's mouth to the requisite size, or hand-broken when tipped on the ground. Long triangular heaps were built, having a series of rough dry-stone flues and chimneys, with brushwood in mounds placed at intervals of 13 feet; these heaps were built parallel to one another, and mostly 98 feet long by 16 feet wide and 8 feet high, containing about 330 tons. The exterior was carefully covered with the finely-riddled smalls or burnt ore that had been previously washed; this enabled the combustion to be controlled, and great care and experience were necessary to ensure a perfect calcination. These heaps were generally burning five, six to eight months, or, as was found by experience, a month for each metre-width of the heap; about 12 per cent. of weight was lost by calcination, and 84 per cent. of the copper made soluble; this permitted the ore to be washed down to 0·20 per cent. of copper, with five consecutive washings. When cold, the heaps were broken open, loaded into wagons, and the material conveyed to the dissolution tanks, into which it was tipped. Here it was saturated with water, the resultant liquor being drawn off when clear; by this process six or seven waters were generally found sufficient to extract the copper, each successive saturation being given greater time; the resultant liquors were passed to a settling-dam, and drawn off as required by the precipitating plant. When the copper in the ore was reduced to about 0·20 per cent. the tanks were cleaned and the washed ore deposited on a site conveniently near, where the residual copper was extracted by further washings; by this process 2½ tons to 3 tons of iron were consumed to produce 1 ton of fine copper, whilst the quality of the precipitate was very poor, often not reaching 60 per cent. of copper.

It was found, however, that by passing the liquors from the lixiviation of the calcined ore over the crude smalls as a filter, a marked change took place, the ferric sulphate being very readily turned to ferrous sulphate, as the copper was taken up from the smalls. This gave an almost ideal liquor for precipitation purposes, and by the development of this process under constant care the consumption of iron in the precipitation of the copper from these liquors was reduced to 1·25 per unit of copper produced, and the quality of the precipitate raised to 80 per cent. Here, then, was a great advance upon the old system; the cost of handling the ore was reduced at once by one-half, and later, this was brought down to a minimum of expense by the burnt heaps being washed *in situ*. The new heaps of crude ore being built upon those that were already washed, there was ultimately only the cost of traction from the pit's mouth to the calcination ground, the making into heaps, calcination and lixiviation, by either washing the burnt ore in dissolution tanks or *in situ*; 35·32 cubic feet of water was used per ton of ore.

The nuisance caused by the calcination of upwards of 200,000 tons of cupriferous pyrites per annum can be readily imagined. Vast clouds of sulphurous-acid gas extended for miles over the country, withering up everything that they touched in their flight; many attempts were therefore made to abolish or at least modify this nuisance, but without result. Many chemical treatments were tried, and were, apart from their partial success, prohibitive in consequence of their expense, the necessary economical treatment of such low-grade ores was a drawback against any complicated treatment. It was, however, ultimately found that an induced natural oxidation of the ores that were specially crushed and tipped in large deposits would, though taking slightly longer to obtain the copper than by direct calcination, give an equally good result, more economically, and at once abolish the noxious fumes. This process has now been working four years, and consists in first crushing the ores to a fineness that 45 per cent. would pass a 1-inch sieve with only 10 per cent. larger than 2 inches; tipping in heaps as large in area as possible, 33 feet in depth, the whole ground occupied by the deposit being laid out in a net-work of dry-stone drains for ventilation and drainage. Occasional small shafts were built up in the heap for purposes of testing the temperature, a record being carefully kept of these readings for reference, as the success of the operation consists in maintaining a regular heat through the deposit. Ores containing so much sulphur oxidize very readily, and the heat generated in the deposit is also influenced by the percentage of copper contained in the ore. By a very careful system of intermittent washing with fresh water, oxidation is encouraged, and, three months after washing operations begin, considerable heat is noticeable in the heap; in six months, the full heat that can be permitted is attained, and from that time forward a check has to be kept on the general temperature. As the copper is washed out, there is a tendency for the heat to subside. As far as experience has indicated, the temperature should not be higher than 120° F., whilst below 70° the rapid extraction of copper is retarded. From ores containing between 1 per cent. to 1·50 per cent. of copper, 45 per cent. of the total can be washed out during the first year, 20 per cent. the second year, and 10 per cent. in the third year. This leaves a small residue, obtainable if desired by light washings. After three years' washings the original tonnage laid down for treatment is reduced by 23 per cent. in consequence of its loss in copper, sulphur, iron, etc. Ores richer in copper will yield a better result, the poorer class slightly worse in consequence of their origin in the lower floors where it is unoxidized and more compact in structure. The top of the heap has a gradient of not less than 1 in 300, which is formed in tipping, and the whole surface is divided into equal areas by shallow drains made with the finer ore, along which the water is conveyed to such parts as are under treatment. The water filters rapidly through the heap, taking up the sulphate of copper produced by oxidation; the whole of the water used in washing, less 19 per cent. absorbed and evaporated, is caught by a main drain which conveys it to a valley filled with what are locally called smalls, given by the screening of the export ore. There being over 750,000 tons in this valley, the area is large and forms a convenient filter for the reductions of the ferric sulphate in the liquors received from the upper heap. Besides the advantages gained in the suppression of the smoke, the ore, after it has been under treatment three years, becomes a valuable asset, its sulphur contents giving it a decided commercial value. The whole process requires considerable care and attention, and is kept under working control both night and day by a number of trained assistants. The liquors obtained by the treatment contain per gallon on an average 210·3 grains of copper, 1225·3 grains of iron as FeO, 171·0 grains as Fe_2O_3, 1296·9 grains of sulphuric acid, 11·22 grains of arsenic, etc.; whilst after passing the filter the amounts are

273·4 grains, 1421·6 grains, 36·5 grains, 1366·9 grains, and 11·22 grains respectively. This liquor is passed on direct to the precipitating plant.

The cupriferous schist extracted from what is known as the Esperanza lode is tipped in a valley below, and in close proximity to the lode. The output amounts to a little over 300,000 tons per annum, and, as a similar extraction has been made for several years, there is now an immense deposit in the lower valley. The material is tipped as it is received from the lode and undergoes no preparation such as the mineral before washing. The most rapid and efficient method of obtaining the copper is to first give a saturation with an acidulated liquor such as is pumped from the mine, or a spent liquor from the precipitating department; and, after two or three months' rest, the heap is in condition for leaching with fresh water. The schist washes freely and there is but little difficulty in reducing its copper contents to 0·10 per cent. The liquors given by the leaching are not strong in ferric sulphate, and therefore a contact of short duration with small crude ore is sufficient to bring them into excellent precipitating condition; these liquors, though seldom rising above 140 grains of copper per gallon, produce, when passed over iron, a very fine class of precipitate, free of arsenic and assaying 87 per cent. of copper.

The quantity of water requisite for the efficient washing of the crude mineral is 440 gallons per ton of ore for the first year, 220 gallons per ton the second year, and 110 gallons for the third year. If the tonnage of ore laid down for treatment each year be more or less constant, it follows that during the third year the largest consumption of water takes place, as not less than 250,000 tons of ore are annually placed under treatment, besides 300,000 tons of cupriferous schist; very large storage reservoirs are required to enable the work to be carried on without intermission during the dry season, which may be considered to fall between the months of June and October. Provision has therefore been made in Tharsis alone for the storage of upwards of 572,000,000 gallons of water brought to the reservoirs by many miles of catch drains. The burning of the cupreous ores for so many years has cleared the country for some distance of vegetation, the ground is therefore in a condition to give a considerable proportion of the rain that falls; many tests have been made and the average over thirteen months was found to be 54 per cent. of the total rainfall.

The evaporation is very great and is found to represent, in the case of the largest reservoir, 96,800,000 gallons per year. During the months of July and August very great heat is experienced, the thermometer occasionally rising to 110° F. in the shade, with 85° for several weeks as a minimum; the healthiest portion of the year, however, corresponds to these hot dry months, and the most unhealthy towards the end of September or the beginning of the first autumn rains.

A Table is given in the Appendix of the rainfall for the last fifteen years and the evaporation for the last seven years.

SAMPLING.

In treating such large quantities of poor ores and schists accurate sampling and analysis form an essential part of the operation. The tendency has always been to err on the higher rather than the lower side; and, under various methods of sampling, designed either with the object or not of counteracting the human error, the same constant tendency is noticeable. It may be safely assumed that no large heap of cupriferous ores has ever given more copper than was originally returned by sampling and analysis; in fact, the whole history of the working of these immense heaps has proved that their copper contents were less than was at first estimated. As the utmost care and vigilance is required, a system has been instituted which is equivalent to mechanical sampling; each wagon at the pit-mouth, as well as on the tipping ground, yields its quota, a method which ensures never less than two returns, and in many cases a third sample is independently taken. The undoubted possibility of error, coupled with the absorption of the ground, etc., makes it impossible to expect that the whole of the copper in the ore laid down for treatment is ultimately obtained.

PRECIPITATION OF THE COPPER.

The general disposition of the ground and its natural fall from around the lodes has been taken advantage of in the arrangement for the supply of fresh water, etc.; the valleys running from the fresh-water reservoir being utilized in the upper portion for the tips of crude ore, below which there are the filter-beds and a series of collecting dams, and, finally, in continuation, the precipitating plant. The most approved method of precipitating the copper from its liquor is by passing it over pig-iron laid in a double layer in canals constructed especially for its reception. These canals, of creosoted timber, are 2 feet 9 inches wide with a depth of 9 inches laid in quadruple, triple or double series; if the fall of the ground permits, about half the total length is laid with a fall of not less than 1 in 200, a quarter of the length following with 1 in 100, and the final quarter with 1 in 50. This arrangement permits of a gradually increasing agitation of the liquors as they advance and become poorer in copper, thereby improving and enriching the precipitate produced. One metre length of canal of the above dimension holds 573 lbs. of iron, and is capable of producing one ton of fine copper per annum. There is, therefore, considerable length of these canals employed at the mines, the total reaching 9,624 lineal yards, the pigs of iron used are 2 feet 6 inches long, weigh 33 lbs, and contain on an average 94 per cent. of iron. The liquors on entering the service contain per gallon on an average 227·8 grains of copper, 1480·5 grains of iron as FeO, 33·65 grains Fe_2O_3, 1380·9 grains of sulphuric acid, and 11·22 grains of arsenic; hundreds of thousands of gallons are passed per day, and, by the affinity which copper has for iron, a rapid galvanic action is set up, the free acid attacking the iron, giving sulphate of iron, and the copper taking the place of the iron in a metallic state. The liquor, when rendered of its copper, passes out of the service with 0·841 grain of copper per gallon, 1811·4 grains of FeO, 1331·9 grains of sulphuric acid, and 5·748 grains of arsenic; the copper is therefore almost all precipitated, the iron increased, and the ferric sulphate turned to ferrous sulphate, at the expense of the metallic iron. A small portion of the free acid is used in the precipitation of the copper and about half the arsenic is precipitated. Each day a certain length of the canals is cleaned, the pig-iron is lifted out, the metallic copper scraped off, and the iron replaced; a " cleaner " can despatch on an average thirty-four lineal yards of canal per day. The precipitate is loaded into wagons, taken to a yard where it is washed with fresh water, being drawn by large hoes against a running stream, the heavier grains and scales reaching the top of the washing canal, the finer grains passing on with the water to a depositing canal where they settle according to gravity, the lighter portion being deposited in a series of tanks adjoining the canal. In this manner an economical classification is made automatically. The next operation consists of throwing the scales into heaps and drying, after which they are ready to be sacked for export; the larger grain precipitate is pressed into cylinders of 22 lbs. weight, and, when dry, also sacked, whilst the poorer class deposited in the tanks is made up into heaps and burnt. By this arrangement the large portion of the arsenic, which is extremely injurious to the quality of the copper, is concentrated in the calcined precipitate. The foregoing ·

classification produces in relative proportion 72 per cent. of scales, 12 per cent. of cylinder, and 16 per cent. of calcined; the scale precipitate containing 92·5 per cent. of copper, the cylinders 78·5 per cent.. and the calcined 48·3 per cent., the average contents being 81 per cent. of copper, with 2·80 per cent. of arsenical ferric oxide, the residue of 16·20 per cent. being graphite from the pig-iron, silica, etc.

The amount of iron consumed in the precipitation of the copper is about 1·25 time that of the copper produced; this is considered a very favourable result, although, according to theory, 56 parts of iron should precipitate 63·5 parts of copper, or in the ratio of about 8 to 9.   This result, is, however, never attained in practice, owing to the ferric salts which are always present, as also the free oxygen which is relatively in large proportion in weak solutions such as are used, and tends to oxidize the iron; the arsenic also requires for its precipitation about 1·5 per unit, whilst the pig-iron itself generally contains not less than 6 per cent. of impurities. The whole of the liquors before entering the cementing plant are measured and analysed, a record being kept of the total amount of copper produced according to the liquors received.   This serves two objects; firstly, the production of the month is known before it is possible to ascertain it by the actual copper produced, as generally three weeks elapse before the precipitate is dry and ready for export; and secondly, should there be more than one source from which the copper is derived, by the aid of the measurements and analysis it can be credited with its due amount. The company produces annually between 10,000 tons and 11,000 tons of fine copper.

RAILWAY AND PIER.

The main establishment at Tharsis and the port of embarkation are connected by a railway of 4-foot gauge, having ordinary web and flanged rails of 60 lbs. per yard spiked to the sleepers.   Six to eight trains run daily according to the requirements of the shipping, each train carrying 100 tons of mineral, whilst an up and down passenger service is also worked in combination with the mineral trains.   Each locomotive weighs 25 tons and the entire journey of 29 miles occupies two hours.   The cost of working the service is $\frac{1}{2}d$. per ton of ore per mile, the line being kept in efficient working state by the employment on the maintenance of one man per mile of road.

The pier is an iron structure, 2,624 feet in length, having at its head ample berth for three vessels; it is provided with turn-tables and steam cranes, the whole being equipped for the loading of 2,500 tons per day if necessary. Near to the pier is a storage-ground capable of holding 30,000 tons of ore, by the assistance of which the loading of the vessels is carried on continuously, there being no necessity to wait for down trains. During any reduction of shipping the excess received from the mines is tipped into the deposit, thus enabling either a large or small tonnage to be supplied to the pier-head for shipment, and a regular train service to be maintained; should also any serious breakdown occur there is always between 20,000 tons and 30,000 tons of ore to draw upon during repairs; contingencies are therefore amply provided against.

LABOUR.

The number of persons employed at the mines, on the railways, pier-head, etc., is, including men, women and boys, about 3,500. The average working day is eight-and-a-half hours, and overtime is of exceptional occurrence. The following are the wages paid for the different classes of labour:—

	s.	d.
Underground miners	3	0
,, fillers	2	6
Opencast miners	2	6
,, fillers	2	2½
Masons	3	0
Carpenters	3	0
Ore washers	2	0
Canal cleaners	2	9½
Mechanics	4	0
Engine drivers	4	0
Blacksmiths and fitters	3	6
Labourers	2	0
Mule drivers	2	6
Boys	1	2½

TREATMENT OF ORE IN ENGLAND.

The exported ore is shipped to Cardiff, Newcastle-on-Tyne, Glasgow and other ports, where it is treated by the alkali makers for the production of sulphuric acid; it is then returned to the works of the Tharsis Company at these ports, where its copper, silver and gold are extracted, the residue being sold as iron ore. The calcined ore, as received from the chemical manufacturers, is first tested for sulphur, which should, for facilitating the operation, exceed that of the copper contents by about ½ per cent.; when less an addition is made of unburnt pyrites. On obtaining the due proportion of sulphur the whole is passed through a crushing-mill, during which operation 14 per cent. of salt is intimately mixed, more salt being added if there is an excess of sulphur above the required proportion. The mixture is then recalcined in muffle furnaces in charges of $3\frac{1}{2}$ tons, which require about ten-and-a-half hours each for the completion of the reaction. The fumes from the furnaces contain copper, silver and gold, and are therefore passed up condensing towers containing coke, through which water is constantly dripping; the fumes are in this way freed of their valuable metals, and the towers are cleaned once every six months, the coke being specially dealt with. The flues also contain a fine powder which is separately dealt with for its metals, being first washed with acidulated water from the condensing towers which extracts the copper, the remaining portion being mixed with the precipitate from the silver settling vats. On the completion of calcination the ore is removed to a series of wooden tanks in which it is washed with (preferably warm) water, occupying about twelve hours, and the final or tenth washing is given with the acidulated water from the condensing towers, which removes the remaining copper so completely that hardly a trace is observable upon the spades which the workmen use in emptying the tanks. The liquors, as they are run off from the washing tanks, are allowed to flow into settling vats, and to others of slightly larger capacity; at the same time, from a specially graduated tank, together with a quantity of fresh water equal to one-tenth of the volume of the copper solution under treatment, an exact amount of a soluble iodide necessary to precipitate the silver present is run in. During this time the liquors are kept constantly stirred to ensure mixture, and are afterwards allowed to settle during forty-eight hours, after which time the supernatant liquors are assayed and run off into the copper precipitating vats, where the whole of the copper is thrown down by iron. The silver precipitating vats are cleaned once a month, and the precipitate collected at the bottom is washed into a vessel prepared for its reception.

The precipitate is composed chiefly of lead sulphate and chloride which is equal to about 40 per cent. of the whole, silver iodide and subsalts of copper. The latter salts are readily removed by washing with acidulated water, and the residue is decomposed by metallic zinc, which results in a precipitate rich in silver, containing a small proportion of gold and zinc iodide, which is again employed for precipitating when its strength in iodine is ascertained, The cost of the iodine is about 9d. per oz., 13 per cent. of which is lost, whilst 70 per cent. of the total silver in the ore is recovered and 40 per cent. of the gold, the cost of the extraction of the gold and silver being between 8*d.* and 9*d.* per ton of ore. Few tests are made during the progress of the work, but the whole operation is most delicate and requires constant skilled attention. The copper precipitate produced, as well as that received from the mines, is smelted and refined in the works.

The Paper is accompanied by two tracings and three photographs, from which the *Figs.* in the text have been prepared.

APPENDIX.

RAINFALL AND EVAPORATION AT THARSIS MINES.

Year.	Rainfall.	Evaporation.
	Inches.	Metre.
1881	36·24	..
1882	14·15	..
1883	25·16	..
1884	25·12	..
1885	41·14	..
1886	21·11	..
1887	31·91	..
1888	36·39	..
1889	16·60	1·57
1890	23·70	1·65
1891	21·08	1·63
1892	35·08	1·50
1893	27·74	1·53
1894	28·01	1·45
1895	46·02	1 35

(*Paper No. 2933.*)

" Tin-Smelting at Pulo Brani, Singapore." [1]

By JOHN McKILLOP and THOMAS FLOWER ELLIS, A.R.S.M.

THE deposits of alluvial tin ore in the Malay States have been for many years, and will probably long continue to be, the chief source of the metal. These deposits, as well as those in the adjoining countries of Siam and Southern China, have, during three centuries, been worked by the Chinese, and to a less extent by the Malays and Siamese.

Until recently the ore was smelted by the Chinese in a most primitive manner; charcoal or half-charred wood being used as the reducing agent in small clay cup-shaped furnaces, with a blast furnished by a sort of air-pump made of wood and worked by hand. This method of smelting is still largely practised, though it is probable that before long it will be abandoned. Its continuance depends partly on the cheapness of Chinese coolie-labour, and partly on the absence of adequate regulations for the preservation of forests in the Malay States. Such regulations are however being now adopted by local governments in the Malay Peninsula. The enactment of these, or any other cause for an increase in the price of charcoal, would undoubtedly render the Chinese tin-smelters unable to compete against the more refined and economical method of smelting with coal and anthracite in reverberatory furnaces.

To the Straits Trading Company belongs the credit of being the first European company to compete successfully against the Chinese in tin-smelting. In 1885–86, one or two agencies were established in the States of Selangor and Sungei Ujong, and, in the teeth of fierce opposition and prejudice, some of the native miners having been induced to sell ore, smelting operations were begun at the abandoned works of the "Shanghai Tin Mining Company of Perak" at Teluk Anson on the Perak River. It was subsequently decided to build works in or near Singapore, as the

---

[1] The discussion upon this communication was taken in conjunction with the two preceding Papers.

experience gained at Teluk Anson showed conclusively that the drawbacks to successful work in such an outlying spot were too serious.

The site chosen at Singapore was that formerly occupied by a graving-dock and accessory works on Pulo Brani, an island lying south of Singapore Island and west of the town. It is reached from the business part of the town by a drive of 3 miles and a ferry of about $\frac{1}{3}$ mile. The island is about 250 acres in extent; and the channels by which ships approach it are fairly easy to an experienced pilot. The chimneys of the works form a conspicuous feature of the view on entering the harbour from the west, and will have been noticed by any one who has visited the capital of the Straits Settlements in the last eight years. Smelting at these new works was begun in December, 1887, with one 2-ton furnace, and has continued ever since. The works rapidly increased in extent, and at the end of five years practically covered 8 acres. They now consist of twelve 4-ton furnaces with accessory plant.

General Arrangement of the Works.—The ground plan of the works is shown in Fig. 1, Plate 3. Everything, with the exception of the European quarters and part of the refinery to be afterwards explained, is on one level—6 feet to 8 feet above high-water. The ground sloped naturally to the sea-front, and a good deal of cutting and filling has been done at various times to level the place. The high ground at the back, 20 feet to 25 feet above the works level, is reserved for European bungalows. Coal-ships and local steamers lie alongside the wharf, and lighters discharge ore, etc., from the dock direct into the ore-room.

The sheds covering the furnaces, machinery and coal-sheds have light iron roofs covered with galvanized iron and carried on iron columns. The store and mixing-room is a brick building about 250 feet long by 50 feet wide. The refinery and metal store are similarly built. The bungalows for Europeans are of wood, surrounded with wide verandahs and carried on brick piers. The huts for the coolies are light wooden buildings carried on wooden posts, covered with the thatch of the attap palm—a style of building suitable to the climate and to the habits of the natives. The blacksmiths', carpenters', and other workshops are wooden sheds covered with attap. The superior native servants, mostly clerks and weighmen, have each a brick house, two storeys high, built in the local style with an air-shaft in the centre.

Buying and Handling Ore.—By far the greater quantity of ore landed on the wharf of the works is bought by the Company's officers at various agencies in the native States. Both at the agencies and at the works the value of the ore is determined by cyanide assay. If it contains much impurity, the sample is first boiled in aqua-regia, and is occasionally vanned. From the appearance and hardness of the assay button obtained, no less than from its weight, the agent fixes the price he will offer. When bought, the ore is sometimes further dressed at the agencies by various devices. The comparatively small quantity to be treated renders any other than manual power impracticable. Hand-jigging and sluicing are the methods usually adopted, yielding ore of high quality and rich tailings. Fortunately natives can be found to work these tailings over again with infinite pains in a "dulang," or wooden dish similar to the Australian miner's dish, but larger and not so deep. The ore is afterwards dried, packed in canvas or jute bags, labelled, and sent down with a guard to the nearest port, where it is shipped direct to Pulo Brani. The ore, when landed, is carried to the store, weighed and stacked under cover by coolies, under the supervision of a weighing-clerk. The assayer then samples and assays each parcel, and his report determines the subsequent treatment. Those lots which need to be roasted are stored in the roasting-house, while the clean ore is emptied into bins in the mixing-rooms. The cost of the bags is a very serious item. When emptied of ore they are taken to a separate room, cleaned, dried, repaired, packed into bundles of one hundred each, and sent back to the agents.

Great care has to be taken in handling the ore. 5·97 cubic feet weigh 1 ton. As it is worth £40 per ton upwards, it can be easily imagined what great loss would accrue from careless handling. Cast-iron floors would undoubtedly be the least wasteful but for the great initial expense. Concrete covered with cement was tried, and did well where there was no wheel traffic ; the barrows, however, broke it up in six months. The best floor tried was made of wooden blocks boiled in tar and arsenic, and laid as close as possible, without other joint than that formed by the excess of tar.

Preparation of Impure Ores.—The production of good marketable tin depends greatly on the quality of the ore smelted. It is true that a great deal can be done to improve bad metal by subsequent refining, but the results are never really satisfactory. The true way to avoid producing tin of inferior quality is to strike at the root of the evil, and eliminate all injurious impurities from the ore before the furnace is charged. The smelter should throw as much of this duty as possible on the miner. At Pulo Brani a sliding scale is used, by means of which the price paid for the ore depends not only on the metallic tin it contains, but also on the nature of the impurities present. The chief of these are mispickel, copper pyrites, and iron pyrites. Wolfram, though never entirely absent, is not present in sufficient quantities to render profitable its extraction as tungstate of soda, by the Oxland process. Its chief effect, as also that of the various siliceous and titaniferous impurities, is to cause loss of tin by increasing the richness of the slags. An incredibly small quantity of arsenic, sulphur, or copper in the ore, is sufficient to render .the tin produced from it useless for all purposes except that of manufacturing inferior solder. At Pulo Brani, any ore containing arsenic or sulphur is thoroughly roasted at least once. The furnace is of the "blind roaster" type, the ore being in a muffle out of direct contact with the fire. The flame from the fire-box passes first between two arches over the bed and then under it to the flue. During the roasting, the ore is rabbled through the charging-doors along the side of the chamber, a suitable flue taking away the gases and fumes evolved. It is found practicable and cheap to roast when necessary in an ordinary smelting-furnace, logs of mangrove wood being used as fuel, and plenty of air being allowed to pass through the doors of the fireplace.

When roasted, the ore, unless of very poor quality, in which case it is treated with tailings, is sluiced by Chinese coolies, and gives "good headings," which can be smelted directly; "coarse tailings," which need to be crushed; and "fine tailings," which are caught in boxes at the tail of the sluices. The "coarse tailings," after being stamped in a 5-head Californian stamp-battery, are again sluiced. The headings therefrom are re-roasted, and treated on a set of six Frue vanners; while the tailings, together with the fine tailings from the first sluicing, are treated separately, being first somewhat concentrated by passing through a set of fixed buddles, then again roasted and passed over the Frue vanners. Some eight sluices are constantly worked. The stamps are also looked after by Chinese coolies, whilst the buddles and Frue vanners are tended by Kling or Madras coast coolies.

Although this is the general procedure, it is varied greatly according to the nature of the ore. Ore containing copper is allowed to stand for a considerable time between the roastings, to weather as much as possible and to allow the copper sulphate to drain off. Both machine and hand-jigs were largely employed at one time, but the latter proved too expensive to be continued, though they gave excellent results so far as purifying the ore was concerned.

System of Labour in Mixing Charges.—It is usual in tin-smelting works for the charges to be mixed by the furnacemen. This is not a good plan under any circumstances, and it is impracticable where the furnaces are worked by Asiatic coolies. At Pulo Brani all work that can possibly be so arranged is paid by piece. The coolies work under the direction of a contractor, subject to a Chinese clerk to whom the manager delivers his orders in writing, and who is responsible for the weighing and mixing.

The manipulation of the ore is divided into three sections—(1) Discharging from a steamer at the wharf or from a lighter in the dock, weighing, storing, emptying into bins or placing the bags at the roaster, or in the concentrating-shed. This is the work of one gang paid at schedule prices. (2) Mixing; this is the work of a second gang, who have to take the ore and other materials, weigh and mix them, and place the charge in a bin in the charging-room, ticketed to show its destination, at a fixed price per ton. (3) The third stage is the work of a distinct set of coolies, who wheel the charges from the charge-bins to the furnace-door, and leave them ready for the furnacemen to put in. Metal from the furnace to the lighter is treated similarly.

An outline of the detail work of charge-mixing is as follows :— The bins each contain ore of a certain assay value; the day's orders contain directions for mixing the charges by taking so much ore from each bin, in order to keep the assay of a charge constant at a given figure. Welsh anthracite is used as a reducing agent; and drosses, sweepings, skimmings, &c., have to be mixed in ore charges in such ratio as to keep them down in quantity and prevent accumulations.

The Smelting Furnaces.—The furnaces, Figs. 2–9, Plate 3, at Pulo Brani are of the ordinary reverberatory type. There have been many alterations in them, however, from the pattern originally erected in 1887. The distinguishing feature of the latest furnace is the water vault. Tin at high temperature becomes very fluid ; and this property, together with its comparatively high specific gravity, renders it a most difficult matter to prevent leakage. After many trials and attempts to entirely prevent leakage through the bed, all of which failed, it was decided to regulate the leaks rather than to try to prevent them. The evil of these leaks is not absolute loss of metal, but trouble and difficulty in recovering it. Tin melts at 260° C. The foundations of a furnace, and the ground around it, are at or above this temperature for a distance of some feet. Consequently, any tin that leaks into the vault of an ordinary furnace below the bed, remains liquid, and will slowly but continuously find its way through the cracks of the ground until it reaches a place where the temperature is less than 260° C. The distance tin will travel is incredible to those who have not seen it.   The cost of the periodical recovery of all this metal is very great; for the metal is either in huge lumps of 10 tons or more, or else in fine strings and sheets into which it has been moulded by cracks in the clay.   Sand is said to form an effectual bar to the passage of melted tin.   The experience of the Authors is that at comparatively low temperatures it does act as a check; but at higher temperatures the tin and sand become mixed so completely that separation by heat is very wasteful owing to the oxidation of the metal.   Further, anything siliceous round a tin-furnace should be avoided as far as possible.   It will have to be swept up and treated in a furnace sooner or later to extract the tin; and the more silica it contains, the greater will be the quantity of slag produced, and consequently the greater the loss of tin.

This loss of tin by leakage, with attendant difficulties in re-covery, have been entirely overcome by the introduction of the water-vault, below the bed, containing a depth of 8 feet of water. Any drops of tin are granulated in this water and their further passage is effectually checked.   Once a week the water is pumped out and the granulated metal is recovered.   In every case in the Authors' experience, such explosions as have occurred have been due to deficiency of water.   If care be taken to rabble down any heaps of granulated metal which form below the water, and if the water-level be maintained, no explosion of a serious nature can occur.

The bed and lining are the most important parts of the furnace, and the most difficult to keep in order.   It is necessary to build the furnace in such a way that the bed, lining and roof can each be repaired or replaced without disturbing the other parts.   The bed is of fire-bricks laid on end.   In order to reduce the joints as much as possible, the faces of the bricks are ground true before being laid.   They are laid dry, and forced tight with screw-jacks. The rails which carry the bed lie across the furnace, and are divided in the centre.   Here they are carried by a strong iron rail, while their other ends rest on the inner $4\frac{1}{2}$ inches of the wall of the furnace.   The large rail is carried at each end of the bed on smaller rails built into the brickwork, or by pillars built up from the floor of the vault.   Both methods possess advantages.   The large rail is divided in the middle, and is there carried by a pillar. The bed is laid with a fall of $3\frac{1}{2}$ inches to the tap hole from every part of the furnace.   This fall is secured by placing the rails accurately in position, the bricks following them.   The large rail is first placed accurately along the centre-line of the furnace, with a fall of 1¾ inch from the front door and bridge. The 4½-inch work which carries the small rail round the charging-door side is levelled; while that round the tap-hole side is finished with a fall of 3¼ inches from the bridge and front door to the tap-hole. The cross rails can then be placed in position and the bricks laid. Sometimes the large rail, instead of being divided at the middle, is merely heated and bent. This is very troublesome and has no advantages over the method of dividing the rail. When a bed is worn out, it can be quickly removed by knocking down the centre pillar, when the whole collapses. The courses of bricks in the bed are laid across the furnace, beginning at the bridge. One course is laid at a time, and is carefully keyed up while the screw-jack is on. The bricks are all gauged for each course, $\frac{1}{16}$ inch excess or defect on the width (4½ inches) being rejected. When the bed is complete it is grouted with fire-clay cream, dried carefully and heated. The first charge is cast iron, which, when melted, forms an excellent grout and binds everything firmly together.

The lining rises from the red brickwork behind the bed. The end brick of each course of the bed abuts on the lining, which must therefore have a true face and the smallest possible joints. The lining is all in headers. Where it meets the roof it is finished off by a course of three-corner end-splayed bricks. The roof, instead of springing from the lining, springs from the outside work of red brick. As this is only 1½ brick thick, the thrust of the roof is taken by T-iron, built in and supported by the vertical girders which bind the furnace. The thrust of the bed of the furnace is taken in the same way by T-iron built into the brick-work. The bridge is built with as much care as the bed. It cannot, however, be kept tight, and is therefore built with a cavity which is continuous through the outside work. In this way any slag which leaks through and sets can be knocked off with a steel bar. Tin which leaks through the bridge falls into the water in the vault. The doors of the fireplace are in the back wall opposite to the bridge and high up. The fireplace is easily filled through these doors, and the fire-rabble is rarely needed. The coal lies at its proper angle of repose from the roof above the fire-doors down to the bridge, and there are no empty corners possible. Winding is done through a cast-iron winding-plate placed below the fire-doors. The lower row of holes in the plate is about 9 inches above the bars. This form of fire is very easily worked. The 'flote' or pot into which the metal runs when the furnace is tapped is a wrought-iron or steel tank lined on the bottom with 9-inch, and round the sides with 4½-inch fire-brick. These leak in spite of all efforts to keep them tight, and the water-vault has been extended under them with good results. In case of any hot material getting through the bed with a rush, two pipes, 18 inches in diameter, are built into the thick corner of the furnace in order that the steam may escape freely.

The working of the furnace is as follows. Suppose a charge has just been drawn. The doors are open and the bed and walls are inspected. If much worn and eaten away they are ' fettled.' A mixture of bauxite and fireclay moistened with water is put on the worn place with a paddle, and is rammed home with the head of a rabble. When all the bad places are covered, the doors are lowered and the fettling is " glazed " by hard firing for about an hour. This fettling should be required only once a week, in addition to that given when the furnace is overhauled on Sundays. When the furnace is hot and ready, the doors are opened, the damper is closed, and the charge is thrown on the bed through the side doors, while the leading coolie levels it with a rabble through the front door. The charge being all in, the doors are closed and the fire is made up as large as possible. Meanwhile the charge-wheelers bring out the next charge and tip it under the charging-doors, and one of the four furnace coolies turns it into two heaps, one under each door. The leading coolie then turns his attention to the slag-beds, and prepares them to receive the slag from the charge. As soon as he sees that a fresh fire is needed, he calls the European smelter, who, after inspection through the peep-hole in ·the front door, decides whether to put on another fire or to rabble the charge. With good coal the first fire should last two hours or longer. This gives the charge a proper start, after which it may be rabbled. It should be liquid near the bridge, and only moderately thick at the tap-hole, where it is deepest, and towards the front door, and frothing freely all over. A good rabbling at this stage should free it from the bed, and mix it thoroughly. The fire is again made up as full as possible, and when it has burned clear the rabble is again used to ensure that everything is loose from the bed. At this time the surface of the charge in the furnace should be resplendent and free from floating lumps and patches. If so, the door is closed, another fire is put on, and the tapping-bar is withdrawn. A stream of tin ¾ inch in diameter escapes and falls into the flote. At this rate it requires about forty minutes for all the tin to drain out, leaving only liquid slag in the furnace. When it has been ascertained that all the tin is out, the tapping-bar is again inserted, and the channel from the tap-hole is altered to deliver over the slag-beds. The whole of the clay stopping of the tap-hole is removed, and the slag, rushing out, fills the slag-beds. The tap-hole is then closed, and the furnace is recharged.

Metallurgical Processes.—The metallurgical processes employed may be conveniently considered in four parts.

(A) Smelting ore, with the production of "rich" slag and "ore metal"; (B) Smelting rich slag, with the production of "poor" slag and "rough metal"; (C) Treatment of poor slag containing tin as prill; and (D) Refining the metallic products of (A), (B), and (C).

(A) A charge is made up by mixing ore with between 13 per cent. and 15 per cent. of culm or anthracite, and about 3 per cent. of refinery dross. If the quantity smelted at one time is 4 tons, the composition of the charge would be somewhat as under :—

—	Poor Ores, 65 per Cent. and upwards.	Rich Ores, 71 per Cent. and over.
	Cwt.	Cwt.
Ore	80·0	80·0
Culm	10·4	12·0
Dross	2·4	2·4

The time required for a charge should be seven hours or eight hours with good coal and labour, but sometimes longer periods are required. From a charge of such composition there should be obtained 45 cwt. to 48 cwt. of metal containing about 99·5 per cent. of tin, and 29 cwt. to 30 cwt. of rich slag containing 30 per cent. to 40 per cent. of tin.

The metal from these charges is hard, brittle, and dull in colour; it is rather greyer than refined tin, and if poured hot, it may be covered with beautiful iridescent films of oxide.

The slags produced at this stage, distinguished as "rich slags," are variable in appearance—sometimes dark brown and highly crystalline, and sometimes quite black and glassy. In thin sections they show under the microscope a yellow matrix with numerous black crystallites. Reflected light shows tin prills and at times a brown substance, probably ore that has only been fused. Their composition varies widely. Specimens examined contained 35 per cent. of tin, 15 per cent. of silicon, 18 per cent. of aluminium, and 9 per cent. of iron, in addition to manganese, titanium, lime and magnesia.

(B) There are many ways of smelting rich slag. They may be tabulated as follows :—(1) Smelting rich slag with excess of scrap-iron and cu'm to produce 'hardhead' and poor slag unfit for further use. This method alone is not desirable, as the hardhead produced (an alloy of iron and tin very difficult to separate) is difficult to treat further. (2) Smelting rich slag with culm and sufficient iron to decompose the tin silicate. The difficulty with this method is the danger that the slag produced should be too rich to throw away; but the metal (termed 'rough metal') is fairly

DIAGRAM SHOWING THE PROCESSES OF SMELTING.

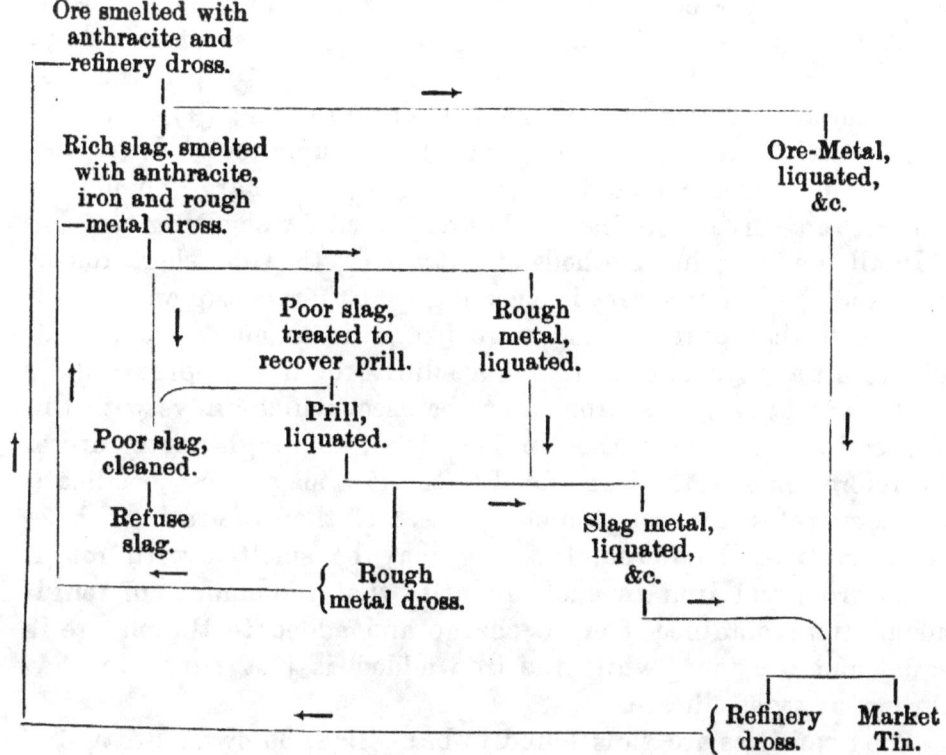

soft and not difficult to refine. (3) A combination of the first two methods, by which hardhead is first produced by excess of iron, and is subsequently smelted with more rich slag; the result being that the rough metal is fairly easy to refine and the slags are sufficiently poor to be rejected. The process, however, requires great care both in mixing and smelting.

In refining the metal from slag, whatever process is employed, a heavy black dross always remains, containing iron oxide and tin oxide. This is called "rough metal dross," and has to be worked up continuously. The two following methods of working slag are therefore also used :—(4) Rich slag is smelted with culm and rough metal dross, producing rough metal and poor slag—a process which also requires great care—or (5) Rough metal dross is used to replace hardhead in process (B) (3). The greatest objection to all these and similar processes is their intermittent nature. Continuous and regular work is most desirable in metallurgical as well as in other industrial processes.

The mixings required for the above processes are respectively as follows:—(1) 1½ ton of rich slag per charge mixed with 20 per cent. to 27 per cent. of iron and 23 per cent. to 26 per cent. of culm, according to the richness of the slag; (2) 1½ ton of rich slag with 16 per cent. to 20 per cent. of iron, and 20 per cent. to 25 per cent. of culm, according to the richness of the slag; (3) 1½ ton of rich slag with 62 per cent. to 70 per cent. of hardhead and 24 per cent. to 25 per cent. of culm; (4) 1½ ton of slag with 62 per cent. to 70 per cent. of rough metal dross and 20 per cent. to 24 per cent. of culm. A 1½-ton charge should be smelted in a furnace similar to that described for smelting ore.

In all the foregoing methods of extracting tin from slag, iron is used, and there is difficulty in ensuring that the exact quantity may be used so that pure tin and pure iron silicate may be obtained. The result always is, that to get a silicate of iron approximately free from tin, excess of iron has to be used, which alloys with the tin, giving rise, when the tin is refined, to rough metal dross. Therefore some of the iron added to the rich slag is lost as silicate and some returns as rough metal dross. If then the ratio of these two quantities be found, the slag may be smelted with rough metal dross and iron in such quantity that the amount of rough metal dross obtained from a charge and added to the charge is equal and constant, while the iron added is just equal to that thrown away as silicate.

The quantities are thus found to be:—Slag, 30 cwt.; dross, 6·5 cwt.; iron, 2·75 cwt.; but the slag resulting from this rational composition of the charge was too rich. An addition of rough metal dross led to the disappearance of this trouble, and the proportions adopted were:—Slag, 30 cwt.; dross, 12 cwt.; iron, 2·75 cwt. The excess of rough metal dross can do no harm, as it only circulates unchanged. In addition to these constituents culm has to be added. Slag is ferrous silicate, and a reducing action must take place simultaneously with the replacing action. Coral or lime in any form is also added as a flux to combine with some of the silica present. It is easy to add too much, in which case it combines with the oxide of tin present, and carries it into the slag.

The final composition of the charge would be, therefore, slag, 30 cwt.; dross, 12 cwt.; iron, 2·75 cwt.; culm, 6 cwt.; coral, 2·4 cwt. This method of smelting slags gave satisfaction in every respect. The details of the process are similar to those given under ore smelting. In smelting slag, however, though the furnace and fuel may be the same, a much higher temperature is attained than when ore is smelted, the duty of the fire being less. It is safe to assume that, in ore smelting, the chemical change represented by the equation $SnO_2 + 2C = Sn + 2CO$ takes place to some extent, and this, being an endothermic reaction, may account for the lower temperature of the ore furnace as compared with the slag furnace, other things being constant. Whatever may be the reaction that takes place with slag charges, it is but slightly endothermic compared with the reaction between the ore and carbon.

A slag charge is rabbled three hours after charging, and again an hour later, by which time the charge ought to be well off the bed and the rough metal ready for tapping. The reaction between the slag, iron and culm takes place with considerable violence. When the frothing and bubbling has ceased, the charge is again rabbled, the rough metal is run into the flote, and the slag into moulds. A slag charge should not require longer than between six-and-a-half hours and seven hours.

The products of melting slag are "second," or "poor," slag, and rough metal. The poor slag obtained is hard, black and glassy. In thin sections under the microscope it shows here and there a small amount of yellow matrix; but it seems to consist chiefly of a dark black crystalline body with the crystals closely packed together. It varies greatly in composition, containing 60 per cent. of silica with varying amounts of other bodies that are also found in the rich slags. This slag contains numerous lumps and prills of tin. The lumps are removed by hand-picking after the pigs of slag fall to pieces. The finer prills are recovered as described subsequently, (C). The rough metal is ladled from the flote into moulds and is stirred, while liquid in the moulds, with an iron rod. This stirring is most important, as otherwise the ingots would tend to set in two distinct layers, the lower and larger portion being practically hardhead, an alloy having a high melting-point, while the upper layer would be nearly all tin, holding a little iron in solution. The same result is attained by granulating the metal. In both methods the metal is constrained to set as a uniform alloy, or mixture of two alloys, and is much more capable of economical liquation in the subsequent refining.

The metal is black and dirty in appearance; it is very brittle, and the fracture is steel-grey.

(C) The recovery of the prill may be effected in two ways; (1) by crushing and washing it, as is the practice in Cornwall, and (2) by "running" it in a furnace. The first method was tried at Pulo Brani, but was abandoned, owing to the large amount of the metal reduced to slimes, and rendered difficult of recovery. Slag metal, as pointed out above, is very brittle, and hence it easily forms slimes. The second method is more effective and not more expensive. It consists of re-melting the slags and allowing the metal to sink to the bottom of the liquid charge. It is not necessary to treat all the second slag in this way. By making the first two moulds in the slag-bed deeper than the others, practically all the metal which does not run into the flote can be collected in these moulds; and the slag in the remaining moulds (more than two-thirds of the total quantity) can be thrown away as clean. This leaves less than one-third of the slag to be treated. In this process no chemical change has to be effected; a given mass has merely to be melted and then run off; consequently, the amount of fuel of required is not large.

The charge is rabbled two hours and again three hours after charging, after which the slag only is run off. About three times in the week, the metal which has collected on the bed is tapped off, by giving the furnace another fire after the slag has been tapped; the metal produced being run into sand moulds, and broken up to convenient-sized lumps while still red-hot. This is necessary on account of the extreme toughness of the hardhead when cold; when it is hot there is no trouble in breaking it up. The metal can, however, be treated in the same way as other rough metal, and stirred in the moulds or run into water. A charge would contain, slag, 40 cwt.; culm, 2·5 cwt., and coral, 2·5 cwt. If the slag is free from combined tin, the coral and culm may be omitted. They are only added to effect a further reduction of any combined tin in the slag. Four hours is a full allowance for treating these charges.

(D) Liquation is the method principally adopted for refining at Pulo Brani; but 'boiling' is sometimes resorted to under special circumstances. The ingots, or granulated metal, are piled in a furnace heated to incipient redness, wood being used as fuel. Metal which has not been stirred in the moulds, or which has not been granulated, when thus treated, would leave behind large lumps of hardhead in place of the powdery rough metal dross. The products obtained from liquating rough metal, if properly conducted, are about 90 per cent. of "slag-metal" containing 99·5 per cent. of tin, and 13 per cent. of rough metal dross, containing about 65 per cent. of tin, and 25·5 per cent. of iron, partly as oxides and partly as alloys. This dross is mixed into the slag charges, as has been shown, and circulates in constant quantity.

The metal from this liquation is of about the same composition as 'ore metal,' and its further treatment depends on its destination. For ordinary commercial tin, suitable for making tin plates, solder, or for use in galvanizing, it would be mixed with ore metal, and the two finally refined together. Tin required for making foil or for chemical salts must be very pure and free from iron. It is best in making this quality to avoid any mixture of slag metal, but, if this cannot be avoided, the slag metal must be boiled before it is mixed with ore metal, or the mixture must be boiled after the second liquation.

Boiling is usually performed only on such ore and slag metal as has been derived from ores that have needed roasting and dressing. It consists merely in immersing logs of green wood in the molten metal. The tin is kept just above its solidifying point by a small fire under the kettle. The operation lasts for several hours, until the bubbles of steam from the wood cease to bring scum to the surface of the tin. The wood is then lifted out, and the metal is skimmed, ladled into moulds, and sent to the refinery, the skimmings being added to the slag charges.

The slag metal is then fit to enter the refinery. This building is shown in Figs. 10, Plate 3. The operations performed in it are liquation, followed by 'tossing,' which consists in allowing molten tin to fall from a height into a mass of liquid metal, thereby carrying air into the mass and permitting a certain amount of oxidation. The liquating-furnace is a rectangular chamber with a fire at each end; the smoke leaves by a chimney in the centre of the arch. The ingots are piled on the bed and wood fuel is used. The tin runs continually through the open tap-holes of the refining-furnace into the two kettles A and B (Figs. 10, Plate 3) which are situated in the pouring-room. These two kettles are about 3 feet 6 inches in diameter, and are each capable of holding 7 tons of tin. From these kettles the metal is ladled into one of the larger kettles C, D, E or F, each of which is 8 feet in diameter and holds 30 tons of metal. The kettles A and B are about 2 feet above the others, so the metal in falling into the latter is well stirred and aerated, the average fall during the filling of the larger kettles being not less than 4 feet. In these latter the metal is allowed to stand for twenty-four hours, after which it is skimmed and poured into moulds. When cool, the ingots are weighed, and are then stored ready for sale in the ingot room, whence they can be readily loaded into boats at the small wharf. The tin is kept liquid in the kettles at a temperature of as nearly as possible 260° C. On standing, nearly all the remaining impurities settle out from the tin to the bottom of the kettles, and for this reason the bottom 12 inches or so of metal remaining in the kettles is sent back to the refinery to be liquated again.

After liquation in the refinery furnace, the ingots of metal leave behind on the bed of the furnace a grey, powdery body. This substance is known as 'refinery dross,' and is taken to the mixing-room, where it is added to ore charges. One hundred parts of ore metal will give about 96·5 parts of fine metal and 4·5 parts to 5 parts of dross. Refinery dross is a mixture of the oxides of tin and iron with less easily fusible alloys of the two metals. It contains about 65 per cent. of tin and 11·5 per cent. of iron.

Marketable Tin.—Ingots of good tin should on cooling have a clear, bright surface, with a slight depression on the top, and the crystalline appearance of the surface should show a large pattern. The metal should be soft enough to be marked by the finger-nail, should bend easily, and, if partly cut and then bent, the strained surface should have a smooth, silky lustre, appearing rather as if it had been pulled out than either torn or broken. Impure tin will give the 'cry' of tin on bending; but, on cutting and bending the brittleness of the fracture will increase with the impurity. The latter is the only test used by the buyers in the Straits Settlements. It is unsatisfactory, as the four corners of an ingot can be cut and bent so as to show four distinct qualities. The best test is to roll out a piece of the metal. Inferior tin will then show cracks along the rolled edges, and, if the rolling is fine enough, pin-holes will appear through the foil.

Tin, with as little arsenic as one part in 10,000, will break with a crystalline fracture; four parts in 10,000 will distinctly alter its appearance. The upper surface in an ingot of such tin is pitted, and has a frosted appearance. Small holes occasionally appear on its surface. Iron is always present in metallic tin, and it appears that the presence of this metal is to some extent advantageous. It remains to be discovered whether iron acts merely as a corrective to the other impurities present, or improves the ductility of the metal. It may be that in liquating, or in allowing the tin to stand when liquid for some time, the iron assists the elimination of other impurities. Ore metal comparatively free from iron is improved by being mixed with slag metal relatively richer in iron. A case which seems to point to the same conclusion, came within the notice of the Authors in a very striking manner. A large quantity of highly pure ore was smelted, the resulting tin being kept apart during the refining process, and offered separately for sale. The buyers refused to take it, the metal not satisfying the tests applied to it. Analysis showed it to be quite pure. Accordingly it was re-melted and mixed with some inferior metal, and the resulting mixture was bought directly as tin of the best quality.

Loss of Tin.—From the diagram, p. 69, it will be seen that from, say, 100 parts of ore, containing 70 parts of tin, 56 parts of ore metal are obtained in (A), which is sent to the refinery, and some 36 parts of slag, containing 12 parts to 13 parts of the original 70 parts of tin. As the dross from the refinery is sent back to the ore furnace, it may be considered that in the end the whole 56 parts will be obtained. In (B) and (C) the loss of tin commences, and depends upon the amount of poor slag produced, and its richness in tin. The latter may be taken on the average at 5 per cent., and, as 36 parts of rich slag produce about 27 parts of poor slag, the amount of tin thrown away as poor slag will be 1·35 parts, or nearly 2 per cent. of the tin started with. The loss by splashing and theft will bring it up to somewhat over 2 per cent. of the total amount of tin brought into the works. Considering the various refining processes (D), it is clear that when once the work is continuous no loss of tin would take place there; for the various drosses, as in the case of the refinery dross, are returned to the smelting-furnaces. This applies also in the case of the prill in the poor slag, which is returned from (C) to (B).

Consumption of Iron.—It will be evident that the iron added to the slag charges, viz., 2·75 parts in 30 of slag, or 9·17 per cent., is thrown away in the poor slag. Rich slag is about 36 per cent. of the ore, or 50 per cent. of the tin contained in the ore; therefore the consumption of iron is 4·7 per cent. of the tin obtained.

Consumption of Culm.—This amounts to about 17 per cent. of the ore smelted. If the action of reducing tin ore were exactly represented by the equation $SnO_2 + 2C = 2CO + Sn$, the culm required would be 18 per cent. Anthracite landed in Singapore is expensive, and experience has shown that coal of good quality may be used to replace it. It is necessary, however, to use about 10 per cent. more of the latter. Charcoal may be used, but it is very destructive to the furnace, especially to the flue and 'verb,' or entrance to the flue from the furnace. This is probably due to the potash and soda it contains.

Consumption of Fuel.—The coal used at first was chiefly Australian. Latterly this has been almost entirely replaced by Japanese coal, and still more lately, owing to the rise in price of the latter coal during the war, it has been replaced by coal from Labuan and the Tonkin coal-fields. As the wharves at the works allow ships of 3,000 tons to come alongside, the price of freight is not so high as might be expected, and, except during the wool and wheat seasons, freights are low. The varying cost of coal necessitated a large storage capacity at the works. The coal-sheds are capable of holding about 12,000 tons and even this quantity proved insufficient on one occasion. The consumption estimated on the average of many months' regular work, for all purposes (including boilers, blacksmith, &c.), in smelting ore averaging 68 per cent. net return, is 1·15 ton of coal to 1 ton of ore.

Future Improvements.—Considering the locality of the works described, and the difficulties consequent on the employment of native labour, their organization and management are fairly satisfactory, but numerous improvements in economy by using better mechanical appliances suggest themselves. Owing, however, to the low price of unskilled labour at Singapore, many improvements that would effect economy in Europe are hardly worth introducing in the Straits Settlements. The initial cost also of setting up plant and of repairs is much greater there than in England. The following improvements might be made with advantage. Overhead charging through a hopper would save considerable time with each charge. Allowing the slag to run into slag-trucks would, besides saving labour, prevent the introduction of silica from the sand-moulds into the slag-charges, and the consequent increase of the quantity of poor slags which are thrown away. The flote containing the tin from the ore furnace might be made movable, and means be introduced of automatically casting the ore metal into ingots. The rough metal might be granulated by allowing it to pass from the furnace into a deep well of water, and recovered from it by a cage lifted by a crane. In a granulated form rough metal will liquate excellently, and the ladling into moulds and the stirring are then rendered unnecessary.

A regenerative furnace of the Siemens type was tried, but proved unsuccessful, owing to the impossibility of keeping the chambers free from tin. The cost of repairs and minor alterations inevitable at first were objections which proved too great for its extended use. The ease with which the furnace was worked and controlled, together with the economy in fuel, merited, however, a longer trial.

At various works in England tin slags rejected by the Cornish smelters are treated profitably, the tin being extracted in the form of solder. Lead in some form is, by a process devised by Mr. T. H. Heason, smelted with the tin slag. A sample from works of this kind in Cornwall was found still to contain more than $3\frac{1}{2}$ per cent. of tin. The tin industry in that county seems to be suffering from the lowness of the price of the metal and the increased production in the Straits Settlements. Investigation into the matter might, however, reveal other causes for this depression. A sample of Cornish slag which was sold as road metal was found on analysis to contain more than 15 per cent. of tin.

The Chinese, as has already been mentioned, smelt their ores on an entirely different principle, using small blast-furnaces. Working as they do in a very moist atmosphere, a great deal of the reduction of the cassiterite in their furnaces seems to be effected by what is practically water-gas passing over the heated ore. As is well known, there is an analytical method of reducing cassiterite by heating it in a combustion-furnace and passing hydrogen over the ore. From these considerations, it may be supposed that an entirely new method of smelting cassiterite could be devised; and the future may yet show that the Chinese are working on a more economical principle than Europeans.

In conclusion, the Authors urge strongly on metallurgists how vast is the room for improvements in tin-smelting, perhaps the oldest, and certainly one of the most backward, of English industries.

The Paper is accompanied by six drawings, which have been reproduced in Plate 3.

Discussion.

Sir Benjamin
Baker.   Sir BENJAMIN BAKER, K.C.M.G., President, said the treatment of silver, copper or tin ore did not go on in Westminster, and the question might only remotely affect the majority of the members present, but there were others who were keenly interested in the subject, and the meeting, as representative of the absent members, would, he was sure, join in passing a vote of thanks to the Authors for their very valuable Papers.

Mr. Clemes.   Mr. JOHN H. CLEMES observed that the lixiviation process had suffered a considerable change since its introduction, partly on account of the fall in the price of silver, and partly from another cause which he would mention. At first it encroached on the amalgamation process; but, of late years, the smelting processes had encroached on both the amalgamation and the lixiviation methods of treatment. It was first necessary to increase the output of the roasting furnaces. If that could be done, as had been the case with copper-smelting, it would, with other things, go far to render the process still more useful. Again, when ores were roasted there was a considerable percentage of loss, much of which was not recovered. The flues were made small, the current of gases moved with great rapidity, and the fume had little opportunity to settle. He had been struck with a plan mentioned by Mr. Cowper in a discussion at the Institution in 1893 on smelting processes,[1] and especially with the enormous amount of fume obtained by its means. The principle aimed at was greatly to increase the area of certain of the flues, that the smoke might move slowly, and that the fume might have a chance to settle. That was not done by making the flues at any one point very large, but by putting a number of flues together. No water was used; that was an important point, because the mud or sludge resulting from wetting the fume would be difficult to deal with. For the purpose of making the working of the process more economical, the two objects to be sought were to improve the existing reverberatory furnace by increasing its capacity, and to improve the means by which the material now volatilized and lost could be recovered.

---

[1] Minutes of Proceedings Inst. C.E., vol. cxii. p. 176.

D. C. LE NEVE FOSTER asked the Author of the second Paper D. C. Le Neve Foster. whether *Figs.* 1 and 4, pp. 44 and 48, showed sections through the lode. If so, they were certainly somewhat peculiar. It was curious that the wide lode, *Fig. 1*, should so suddenly run out to what appeared to be a narrow fissure, and end in almost a straight horizontal line. It appeared much more likely that, just as happened at Rio Tinto, the upper part of the lode should continue of equal width although in an altered state. Again, a mass of decomposed ore of equal width coming to the surface might have been expected, rather than the lode ending off horizontally in a straight line, *Figs. 4.* The term "pillar and stall working," which rather applied to bed mining, was apt to engender a certain amount of confusion. He understood the Author to say that the lode was divided into floors 33 feet thick, and that a network of galleries was driven 16 feet high. The process might perhaps be better described by saying that the lode was divided into horizontal slices about 16 feet in thickness, and that the alternate slices were worked away by a network of galleries at right angles to one another. Between each network of galleries a solid floor of ore was left, about equal to the height of the galleries. He thought the other method of mining was incorrectly described as an adaptation of the "long wall" system. The lode was stated to be divided longitudinally at equal distances by diaphragms or walls, but it appeared to be really divided crosswise by those diaphragms or walls. He should be glad to know why it was necessary to adopt that method. Was it not possible to work away the thick lode, or wide vein, by the ordinary filling-up method without making the preliminary cross-cuts, as he should prefer to call them? There was no doubt some reason for it; but it was not stated in the Paper. He thought the method involved the necessity of building a great many more dry walls than would be required if the wide lode were worked away by the ordinary filling-up system. As to a name for the method of working, he thought the process would be better understood if it were called, not an adaptation of the long wall method, but a process of stoping away cross slices of the lode by the overhand system, with complete filling up of the cavities left by the excavation of the ore. In reference to sampling, the Author had said, "A system has been instituted which is equivalent to mechanical sampling; each wagon at the pit-mouth, as well as on the tipping-ground, yields its quota," but he did not say how it yielded its quota. He presumed that from each wagon, as it left the tipping-floor, a sample was taken.

Prof. Roberts-
Austen.

Prof. W. C. ROBERTS-AUSTEN thought the Proceedings of the Institution were rapidly becoming the mines of metallurgical information, for which metallurgists were extremely thankful. The processes described in the first Paper showed how enormously the application of the wet processes had grown in comparatively recent years. The 60-ton vats used in the extraction of silver from its ores in Mexico might be compared with the 500-ton vats in the cyanide process for extracting gold as carried out in Africa. He agreed with the Author that exceedingly accurate results might be obtained by hand-worked as compared with more modern mechanical furnaces, and where labour was so cheap it could hardly be expected that complicated mechanical furnaces would be introduced. He also agreed in the statement (confirming a view he had long entertained) that in processes where cupriferous ores were present in silver treatment, the necessity for employing the Russell process was almost done away with. It was exceedingly interesting to have a recent account of the wet process of extracting copper, but it only showed what excellent results might be obtained by what seemed at first sight to be a cumbersome and barbarous process. It appeared that the introduction of the water-tank, referred to in the third Paper, below the furnace to catch the liquid tin was an excellent and novel feature. He should have been glad if the Authors of this Paper had given other details as to the method of treating slags which were rejected by the Cornish smelters.

Mr. Kitto.

Mr. B. KITTO remarked that the processes described in the first and second Papers, appeared to be successful only when carefully carried out, the products in each case in their different stages being analysed and closely watched. With regard to the third Paper, he thought, as Prof. Roberts-Austen had said, the tank under the furnace was quite new, but he questioned whether it was really necessary. He was under the impression that the amount of tin which got into the ground under the furnaces in the Cornish smelting-houses was very small. It had been stated by the Authors that in Cornwall the slags were sometimes very rich, and one case was mentioned in which slag used as road metal was found on analysis to contain 15 per cent. of tin. He had seen slags from Cornwall containing that amount, but at present it was difficult to find slags so rich. He saw no mention in the Paper of the amount of tin yielded by the analysis of the slags. He had good authority for believing that the primitive method of smelting by the Chinese, although adopted on a small scale, was very effective. He questioned whether the smelting as carried on in the reverberatory furnaces, taking the Mr. Kitto. different scales of operation into account, was really more economical than the old Chinese method.

Mr. A. K. BARNETT, referring to the last Paper, could see no Mr. Barnett. great distinction between the Cornish method of buying and handling the ore and that adopted at Singapore. In the latter case the cyanide method was adopted for the assays, whereas in Cornwall the tin was assayed by smelting a quantity of the ore with culm, which was similar to what was done on a large scale in the furnace; but cyanide was used also as a further guide. As to the preparation of the ore, the Authors wisely said that a smelter should throw as much of that duty as possible on the miner. In Cornwall the ore was, with few exceptions, bought in the state in which it was to be smelted. Occasionally parcels of ore were brought to the furnaces which might require a little washing, but there were not the appliances for what in Cornwall went under the general term of dressing. It should be remembered that the Cornish miner had to deal with a very different class of material from that of Singapore and the Malay Peninsula. Most of the tin obtained there was of a kind which had been worked in Cornwall hundreds of years ago. Deep mining had not been resorted to, and the vast amount of impurities had not been encountered. The average yield of Cornish stone did not exceed 2 per cent., so that 98 per cent. had to be separated before it was brought to the smelter. As to the method of mixing and storing the charges, a great deal must depend on the surrounding circumstances. What was absolutely essential in Singapore might not be necessary in Cornwall. That would apply especially to piece-work. Where a large number of men were employed, piece-work was generally recognised as an economical method of working. In Cornwall the total number of men employed in the largest smelting works was only sixteen night and day, so that no great economy could be effected in labour. The men were not paid by the hour, but in many cases belonged to families that had worked for generations, and took as much interest in what they were doing as the proprietors or managers themselves. There were two shifts of twelve hours, from six in the morning till six at night, and from six at night till six in the morning. During the night, when four furnaces were at work, with two charges in each, the labour of wheeling the charges to the floors, tapping the furnaces, ladling out the metal, removing the slags and everything else, rested on five men. There was one man at each furnace, and another man, locally known as a " tender,"

Mr. Barnett. waiting on the four. As to the water-vault, he had intended to bring from Cornwall, as a practical illustration, a tin dropping, which would have shown that a water-vault was not an absolute necessity, as it might be at Singapore. It was stated in the Paper that a temperature of 260° C., above the melting-point of tin, was found at a depth of 3 feet or 4 feet below the vault of the furnace. In Cornwall, any leakage which came from the bottom of the furnace, instead of sinking into the ground below, really formed a stalagmite by a succession of drops built up. At a temperature of 260° C. it would not be possible to get a stalagmite because the tin would melt and percolate through the ground. In regard to the construction of the furnace, he was reminded of what Dr. Percy had formerly said in relation to a chemical analysis going into seven places of decimals, which he called "an affectation of accuracy." He was inclined to say the same in reference to the details mentioned in the Paper about the squaring of the brick, the use of the screw-jack, and the like. In Cornwall no skilled labourers were employed. The furnaceman built his own furnace, and having examined the bricks to see that they were sound and square, built them into the bed without a jack or anything of the kind. He could rebuild the furnace-bed in two days. In a case of entire rebuilding, including all the interior lining, the bed, the roof, and the fire-place, the men would let the furnace out on Saturday night, get the water poured in during Saturday and Sunday, and by the following Friday night, or Saturday morning at the latest, everything would be ready and the fire lit for the next smelting. To turn out a furnace and rebuild it in one week was, he thought, fairly expeditious work. It was stated in the Paper that the slag taken off at the first operation gave a yield of 30 per cent. or 40 per cent. He had not seen any slag containing so high a percentage. It was stated that on microscopic examination of the ore brown patches were found partly fused. If in the first smelting operation the reduction of the oxide was not complete there was certainly great risk that the unreduced tin-oxide would be coated over by slag, leading to extreme difficulty in reduction at a later stage. He did not see the advantage of large 4-ton furnaces. He did not object to large furnaces, but he should be glad to know why they were so large as the Authors had described. In Cornwall 2-ton furnaces were adopted, which he considered a reasonable size. With the 4-ton furnaces he understood that only three charges could be drawn in twenty-four hours, and eight-hour shifts were therefore necessary. The result would be 12 tons in twenty-four hours. With 2-ton furnaces working four charges Mr. Barnett. there would be 8 tons in twenty-four hours. Larger furnaces would entail different arrangements with regard to the men. Each set of men smelted two charges of ore. He could not see that any great economy in fuel or otherwise would result from changing. With regard to the refining and boiling operation, in Cornwall all the metal was run out into a large kettle by liquation. Green wood was put in, and the gases escaping from it kept up a constant ebullition. That exposed and oxidised the impurities associated with the tin. The top was then skimmed off, and the skimmings were somewhat analogous to the ordinary hardhead. The tin was subjected to the mechanical tests referred to by the Authors, and if satisfactory was ready for the market. He could not agree with the Authors' objection to the Cornish rough mechanical process to determine the commercial quality of the tin, or with their preference for rolling. He knew that the best quality would give a wire edge on rolling, and an inferior quality would give a broken serried edge. Still, his experience differed from that of the Authors in regard to the question of copper. It was often found in Cornwall that traces of copper did not interfere with solder manufacture. He had never had a fair coppery tin returned as objectionable from the solder maker; it made a high-class solder tin. Arsenic was much objected to and should be got rid of. With regard to the use of the Siemens regenerator furnace, he saw the difficulty attaching to it in consequence of its blocking the air-chambers and other things of that kind. He still believed in gas-fired furnaces, and should be glad to see them tried at Singapore, where the fuel was expensive and the plant was much larger; he should advise the use of the Wilson gas-producer, to see if a gas-fire could not be obtained apart from the ordinary furnace. The percentage contained by Cornish slag had been mentioned by the Authors, but not the percentage in the works they were dealing with. As to slag being thrown away as road metal which contained 15 per cent., Cornish smelters had not so disposed of their refuse slags for fifteen or twenty years. They found a ready market for them in Wales. One reason why they did not smelt their own refuse slags, was that they had to buy their fuel; and another was that they would have to enter into competition with the very people who bought the slags for the next product which was desirable—lead refuse of some kind. They had an advantage because they could buy the various refuses from the Welsh tin-platers. Having that advantage they were able to work the Cornish

Mr. Barnett. slag more economically than it could be worked in Cornwall. He remembered an axiom of Dr. Percy, "The end of all metallurgical operations is the balance-sheet." After referring to the Chinese method, the Authors stated: "The enactment of these (regulations), or any other cause for an increase in the price of charcoal, would undoubtedly render the Chinese tin-smelters unable to compete against the more refined and economical method of smelting with coal and anthracite in reverberatory furnaces." He would call attention specially to the words "more refined and economical method." Yet in a later paragraph the Authors referred to the Chinese smelting their ores on an entirely different principle, adding: "From these considerations, it may be supposed that an entirely new method of smelting cassiterite could be devised; and the future may yet show that the Chinese are working on a more economical principle than Europeans." Certainly those two paragraphs appeared contradictory. Perhaps the Authors meant that while the Chinese principle was right, their practice was wrong, and that if the Chinese principle could be put into better practice, an improvement might be effected. It was further stated by the Authors that the tin industry in Cornwall seemed to be suffering from the lowness of the price of the metal and the increased production in the Straits Settlements. There was one point which they had overlooked, and it was a most important factor. In Cornwall tin was sold for £60 per ton, and for that labour and everything had to be provided. The tin was exported from the Straits to England at £60 per ton, and sixty golden sovereigns (or at least the portion necessary for their work, minus profit) could be changed into Mexican dollars equal to about £100. He thoroughly agreed with the Authors in urging strongly "how vast is the room for improvements in tin-smelting, perhaps the oldest, and certainly one of the most backward, of English industries." Although tin containing a minute quantity of arsenic would "cry," it would require a very large amount of impurity to effect such a result.

Mr. Blount.   Mr. B. BLOUNT observed that an interesting statement had been made in the first Paper tending to show how entirely tradition might rule, even in enterprises which depended primarily upon novel principles. Calcium hyposulphite, and similarly calcium sulphide, had been used almost exclusively in the treatment of silver ore, apparently for no reason except that lime was on the spot. On investigating the question of cost, it was found that the cost of sodium salts was not much greater and presented certain advantages. Seeing that crystallized sodium hyposulphite contained a large percentage of water, a certain advantage would be gained Mr. Blount. (in saving of freight) by sending the salt out dry, if that was feasible. It was stated by the Author that, "The calcium sulphide employed for precipitating the silver always contains some calcium hyposulphite, which partly accounts for the increase in the volume of the stock of solvent." With that view he was in accord, but he did not think the reason for it had been properly expounded. It was further stated, "Flowers of sulphur and caustic lime are boiled together with such an addition of water that the resulting solution of calcium polysulphide marks 8° to 10° Baumé." But besides calcium polysulphide, calcium hyposulphite would also necessarily result. The one could not be obtained without the other. There was a certain amount of oxygen from the lime to be disposed of, and it appeared in the form of hyposulphite. With reference to the third Paper, he was particularly impressed with the Authors' statement as to the extreme liquidity of molten tin; and the device which he had adopted, and which had been extolled by Prof. Roberts-Austen, was certainly a reasonable one. He dissented from the view of Mr. Barnett, who apparently upheld the Cornish practice because it had been carried on to the third and fourth generation. He thought it should have a more valid basis than that. The chemical change involved in the reduction of tin ore was stated to be represented by the equation $Sn O_2 + 2 C = 2 CO + Sn$, being regarded as endothermic. It seemed uncertain that the reaction was endothermic at the temperature prevailing in the furnace. It might also be contended that the main reaction took place with the oxidation of CO to CO_2, in which case the reaction would be exothermic. With regard to refined tin, many interesting points had been set forth. It was said that buyers were accustomed to choose their tin by methods which, with the most lenient views, must be denominated crude. An analysis would tell all that was wanted. It was stated that tin with as little as 1 part in 10,000 of arsenic would have a crystalline fracture. He could corroborate the statement in the case of antimony, for which a very small amount of tin had a profound effect. Alluding to the Chinese smelting of tin, it was stated in the Paper that the men worked in a very moist atmosphere. From this it was deduced that they used water-gas, although they did not know it; certainly, if that was the case, the atmosphere must be abnormally moist.

Mr. CLEMES, in reply, believed the presence of hyposulphite in Mr. Clemes. the calcium-sulphide was distinctly indicated in the Paper. It was a necessary result of the manufacture. He quite agreed,

Mr. Clemes. however, with Mr. Blount's statement as to the probable superiority of sodium salts, which he thought was borne out in the Paper. He had attributed a good deal of importance to the question of temperature in roasting, but the necessity of carrying out the roasting in a highly oxidizing atmosphere might also be emphasized. Trouble had sometimes arisen in the working of one or two types of automatic furnaces from the lack of means to fulfil this condition; sufficient air could not be introduced at the right time or in the right place, and, consequently, the reactions to which he had referred occurred afterwards. Cases had been observed in which pulp roasted under this unfavourable condition had, immediately after being dropped from the furnace, yielded satisfactory results to the usual tests, 90 per cent. or more of its contained silver being extracted by a hyposulphite solution. But after being sprinkled, allowed to lie on the cooling-floor, and leached with water, it was found that the amount of silver soluble in "hypo" had greatly decreased, a number of tests showing 80 per cent. and upwards, and a few as little as 70 per cent. of the contained silver to be extractable by such solutions.

Mr. Courtney.   Mr. COURTNEY stated that a perfectly correct representation of the sections of the lodes was given by *Figs. 1* and *4*; there was no mass of decomposed ore rising to the surface, but only as indicated in *Fig. 1*, and one of the interesting and peculiar features of the lodes was that both in length and width they were remarkably level on the top, there being in some cases a slight undulation on the surface. *Fig. 4* showed by no means an unusual form for the lodes to take, but generally the larger and wider masses were of such great and unworkable depth that it was not known exactly what form they assumed at the lower levels. In one case a diamond bore had passed through one of the lodes at 590 feet depth from the surface, or 240 feet below the lowest possible workings, whilst at the 350 feet depth the lode had not decreased in width, and from the general indications there was no reason to believe that in this case there was any diminution of width at the lowest point touched by the bore. The only justification for the use of the term long wall was that the system employed enabled the whole mass of the ore to be removed, and to carry out this method the lode was cross-cut at equal distances, and at right angles to its major axis, the space being filled with dry-stone walling and waste. This method could only be successfully employed where the lode was of moderate width, as in the case given, and the walling became necessary to resist the enormous pressure and consequent creep of the decomposed foot- and hanging-walls, which filling alone would not prevent. The lode was attacked in several places at the same time, and if carefully arranged any intervening slices of ore that were designedly left between the diaphragms could be removed, and waste only inserted as the ore was extracted. The samples had been taken by a sampler, who was supplied with a shovel and hammer combined, of about 20 inches in length, and was used as a gauge. The shovel was placed against the end of the wagon and the hammer directed towards the centre; whatever piece of mineral it touched was taken, thereby obviating any possibility of choice. These samples were placed in a box, and at the end of the day's work sent to the mill for the necessary mixing and reduction in bulk. The variations in the percentage of copper between the samples taken at the pit-mouth and local treatment ground or the port of embarkation under this system were very trifling.

Mr. McKILLOP observed that there was practically no difference between Cornish so-called "glass" slag and Pulo Brani "first" or "rich" slag. The one might be richer or poorer in tin than the other, but as metallurgical products they differed only in quality, not in nature. The treatment which he would have adopted to extract the metal from them would be that given in the Paper, with a modification in the amount of iron added; obviously, a lower percentage of tin required a smaller quantity of iron to displace it. The quantity of tin which penetrated into the ground under and round a furnace was very large. He believed that on removing a furnace in Cornish smelting-houses the accumulations discovered were considerable, and were regarded as a material assistance towards the cost of renewals. It was noteworthy that ore was bought in Cornwall on the anthracite assay which could not yield as high a result as the cyanide assay. From this it followed that an examination of the books and accounts of a smelting-house would not necessarily reveal a deficiency of tin, when there were considerable accumulations below the furnaces, and the quantity obtained and sold might agree closely with the quantity estimated by the assays. When the tin was actually recovered on rebuilding, its value was so much " found money," and might be regarded as a source of loss to the miners, not as a loss recovered by the smelters. The stalagmitic formations referred to were common in all tin furnaces, but a change of wind or the closing-up of any air-hole in the vault would cause them all to disappear in an hour or two. In Australia the usual method of recovering the "tin droppings" or "candlesticks," as they were called, was to stop all the air-holes in the vault, when they melted and ran down to a small basin

Mr. McKillop. placed conveniently for ladling from the outside. It was while the tin was in this melted condition that it leaked away, and continual though small increments made up in a few months a large quantity. Perhaps there was an average higher temperature under and round the Straits 4-ton furnace than under and round the Cornish 2-ton one, but this did not prove the non-existence of the evil in Cornwall. The average percentage of tin in rejected slags at Pulo Brani was given at p. 75 as 5 per cent. This was an average figure, the results of many months' assaying by Mr. Ellis. There was variation normally between 9 per cent. and $1\frac{1}{4}$ per cent. It was not found economical to work below 5 per cent., and anything above that was admitted to be bad work. With cheaper coal and more reliable labour there was no reason why 3 per cent. should not be the regular average, and it might be possible to work at an even lower figure, as was regularly done in German works. It was sufficient to admit that Straits ores were very largely "stream" ores. Still, "lode" ores were coming into the market, and would increase in the future. At the same time it must be remembered that stream ores were not all pure, or even approximately so. Many months' assay returns showed $70·3$ per cent. as the average tin content. On the question of labour the Authors were of opinion that one establishment on the lines of Pulo Brani would smelt all the Cornish ores at less labour-cost per ton of metal than now obtained in Cornwall. Unfortunately this must remain a matter of opinion only; there was no comparison of a satisfactory nature possible. Fire-bricks reached Singapore after considerable knocking about in a steamer. In Cornwall they required much less handling in transit. He had known actual breakage to reach 23 per cent., and this would be an indication of the state of a shipment in the worst case. The average might be taken at 8 per cent. or 10 per cent. Chipping went *pari passu* with breakage, and hence the greater care that seemed to be required with bricks used for beds at Pulo Brani than in Cornwall. The 2-ton furnace was obsolete everywhere but in Cornwall, and even there it was rumoured that 3-ton furnaces were being introduced. At Pulo Brani the first ones built were of the 2-ton size, but increased capacity was found by trial to be accompanied with greater economy. The limit was reached when the furnace tools became too large for convenient handling. Any further change in size would have to be accompanied by change in principle. It was worthy of note that the gas furnace that was tried smelted a charge of 6 tons in a very satisfactory manner as far as working the charge was concerned. The air and other passages were not blocked as much as was the case in steel furnaces which Mr. McKillop. the Authors had seen in the north of England, Scotland, and Belgium. The gas regenerative furnace had been successfully applied in spite of *a priori* apparent objections; the point to note was that the regenerator chambers must not be placed below the hearth for reasons which it was the main object of this Paper to emphasize. Straits tin had to stand a higher test than any other. The best of it was used by foil-makers on the continent of Europe, and for this purpose the usual bending test was valueless. It might also be pointed out that a large quantity of Straits tin went to Cornwall to be mixed with Cornish tin, after which it was resold as "English." This portion also required to be of good quality. Any one who had worked a furnace could not have failed to notice that the temperature rose with extreme rapidity at the conclusion of the action in any charge. This indicated that there was an absorption of heat during the reduction of $Sn\,O_2$ by carbon, but this was not the only possible explanation of the phenomenon. The Chinese method was a most interesting one, and would repay careful investigation. No one could possibly contend that the method was an economical one. The clouds of tin fume leaving the top of the furnace, the thick deposit of tin oxide which covered everything in the smelting-house, the price of the charcoal required, which was three or four times as dear as coal, the small production per man, all pointed to the wastefulness of the method. The principle of water-gas reduction, on the other hand, was capable of economical application, and would have to be made use of in the future.

Correspondence.

Mr. A. H. BROMLY, of Nankan, Upper Burmah, asked, with Mr. Bromly. regard to the first Paper, for further information as to the output, or preferably the duty per effective HP. per hour, of the 1,000-lb. stamps when used for dry-crushing. The duty of the stamps when working with a coarse screen followed by sieving and return of over-tails to the mortar-boxes would also be interesting. Such data would enable a valuable comparison to be made with stamps when crushing wet. Under the latter conditions the duty varied accordingly as the stamp was run with a view to large crushing capacity, or mainly as an amalgamating apparatus. A wet-stamping 800-lb. mill, with 90 drops per minute, a lift of 8 inches, and an average depth of discharge of 8 inches, had a duty of about

Mr. Bromly. 1·18 cwt. per effective HP. per hour when crushing hard quartz through a sieve of 40 meshes to the lineal inch. This mill was run for amalgamation rather than large output. Upon the Rand, where the conditions to be fulfilled almost exactly corresponded with those prevailing in the lixiviation process, the duty of wet-crushing stamps was between 2·5 cwt. and 3·0 cwt. per effective HP. per hour. The reduction of duty due to crushing dry had been stated to be as much as 75 per cent., and it would be interesting if the Author could supply figures bearing upon this point, especially as it was generally assumed that in this connection a well-designed roll-plant was a superior and more economical method of reduction. Rock-breakers being common to both systems might be neglected. Referring to the question of fine dust or "slimes" as affecting lixiviation, if the Author would furnish figures as to the condition of the products from mills running under the two systems they would be extremely valuable as establishing a record of a good leachable product upon such ores. It was stated that the rate of filtration was nearly doubled in one case by coarse-crushing and after-sieving. He found the 800-lb. mill above-mentioned to give a pulp with between 58 per cent. and 64 per cent. by weight through a 120-mesh sieve and when crushing through one of 40-mesh, the ore carrying about 5 per cent. of sulphides. His experience with rolls crushing hard quartz through 30-mesh sieves showed that between 15 per cent. and 35 per cent. passed 120-mesh, the percentage varying according as the plant was being crowded, or otherwise. Gruson ball mills crushing quartz through 70-mesh gave 5 per cent. through 120-mesh. It was not stated in the Paper whether the fine dust exhausted from the mortars required separate treatment, or was merely mixed with the remainder of the pulp. He would be glad also to know what was the matrix generally of the ores under consideration, this factor being necessarily most important in the construction of a reduction-plant.

Mr. Brough.   Mr. BENNETT H. BROUGH pointed out that, in his description of the geology of the Tharsis district, Mr. Courtney had omitted to refer to the controversy that had long existed as to the true nature of the deposits. As far back as 1876 Mr. F. Roemer [1] had enunciated the view that they were of sedimentary origin and strictly conformable to the surrounding beds of slate. On the other hand, Mr. J. Gonzalo y Tarin,[2] in 1887, had announced his adherence to

---

[1] "Zeitschrift der Deutschen geologischen Gesellschaft," 1876.
[2] "Descripción física, geológica y minera de la provincia de Huelva" (Memorias dela comisión del mapa geológico de España). 1887–1888.

the view that the deposits were veins or lodes formed by subsequent Mr. Brough. infiltration of ore between the surrounding beds, and that view was shared by Mr. de Launay[1] in 1889. The matter had, however, been definitely settled by Professor F. Klockmann,[2] of Clausthal, who, in 1894, brought forward a series of arguments proving the bedded nature of the deposits. The age of the surrounding rocks, which the Author stated to be either Silurian or Devonian, was also a matter of controversy. They were considered by Mr. Roemer to be of Lower Carboniferous age, but the results of an examination of the fossils obtained by Mr. J. H. Collins[3] in 1885 clearly proved them to be of the Upper Devonian period.

Mr. BERNARD DAWSON thought it probable that a regenerative gas Mr. Dawson. furnace might be designed for tin smelting, in which it would be impossible for metallic tin to reach the regenerator chambers, and in which the cost of repairs and renewals, and the time lost while they were being carried out, might be reduced to a minimum. Furnaces of this character were in use for smelting nickel and other ores, and had effected great economy both in the price and in the weight of the fuel formerly used in the reverberatory furnaces they had successfully replaced. Without some knowledge of the design and construction of the gas furnaces mentioned by Messrs. McKillop and Ellis, it was impossible to determine whether they were fitted for their purpose; but that furnaces could be built to do this work successfully, and with some saving in the labour of charging and manipulation of the ores and the finished product, was without doubt.

Prof. HENRY LOUIS had, during a three years' residence at Singa- Prof. Louis. pore, had frequent opportunities, by the courtesy of Mr. McKillop, of studying his process of tin-smelting. He had no doubt that the mixed carbon and iron-reduction method was decidedly superior to what might be termed the Cornish or simple carbon-reduction method. Cleaner slags were obtained by the former, and the consumption of fuel was necessarily less. As Messrs. McKillop and Ellis had pointed out, the direct carbon-reduction was a strongly endothermic reaction, whilst the iron reduction was only slightly endo-, if not actually exo-thermic. Now in the Pulo Brani process, only 80 per cent. of the tin was extracted by the former method, leaving 20 per cent. to be dealt with by the cheaper method as regarded fuel.

---

[1] "Mémoire sur l'industrie du cuivre dans la region d'Huelva," Annales des Mines, vol. xvi. 1889, p. 427.

[2] Sitzungsberichte der königlich preussischen Akademie der Wissenschaften zu Berlin, vol. xlvi. p. 1173.

[3] Quarterly Journal of the Geological Society, vol. xli. p. 245.

Prof. Louis. The first was the great point to be looked to; in smelting a metal worth £60 or more per ton, consumption of scrap iron or even of fuel was a small matter compared to clean slags. He noticed the tin contents of the slags were stated in the Paper at about 5 per cent., but his impression was that a good deal of slag was produced that was much cleaner than that, and that volatilization and soakage into the ground accounted for a good deal of the loss; in volatilization he included ore mechanically carried off by the furnace as well as metal actually vaporised. The loss, however, under these heads was not referred to in the Paper.

With the most of the Authors' views he was in entire agreement; indeed, he thought the process described was a distinct step in advance in the metallurgy of tin. With regard to future improvements, it might, however, be possible to suggest a few that the Authors had not referred to, especially with the object of saving fuel, which often meant also saving of time. Thus he should like to know whether there was any good reason why, in a properly arranged plant, the rich slag should not be run direct into a specially designed slag-furnace. He would point out the immense improvements that had been recently introduced into the metallurgy of copper in America, by treating direct in the molten state products that used to be at one time granulated and subsequently remelted. The time and fuel that might be saved by such direct treatment of molten slag were obviously considerable.

He did not entirely agree with the Authors' views on the Chinese method of tin-smelting, which he had had repeated opportunities of studying. In the first place, the ore treated by Chinamen was very pure; almost the only impurities in it consisted of ilmenite and magnetite. Wolfram was conspicuous by its absence, and this, together with the various forms of pyrites, was to be found principally in the lode tin of Pahang (which was not worked by the Chinese), and in certain districts of Perak. As far as he had seen, the tin from Selangor and Hulu Pahang was extremely pure. The Chinese used two forms of furnace, a draught-furnace and a blast-furnace; their construction was similar, consisting of a short cylindrical or slightly conical stack made of clay, kept in place by bamboo poles and hoops; the interior consisted of a crucible between 9 inches and 12 inches in diameter, cylindrical at the bottom for more than a foot, with a conical stack 2 feet to 2 feet 6 inches high, opening outwards to a diameter of about 2 feet or more at the top. There was a small tap-hole in front, and an opening at the back that admitted a clay tuyere about 1½ inch in inside diameter, or, in the other type of Prof. Louis. furnace, admitted a couple of short clay pipes about 3 inches in diameter. The draught-furnace was preferred, but could only be used with first rate charcoal. The furnace, *Figs. 1*, was a mass of clay with bamboo stakes driven into the ground around it, and bamboo hoops to hold it together. The actual furnace, or crucible, was at E; A was a moulded cylindrical clay tuyere between 5 inches and 6 inches in diameter, 2½ inches bore, and about 22 inches long. The bamboo blast-pipe was shown at F, which conducted the blast from what was practically a double-acting blowing cylinder made of a hollow tree-trunk, 12½ inches in diameter and 10 feet long,

Figs. 1.

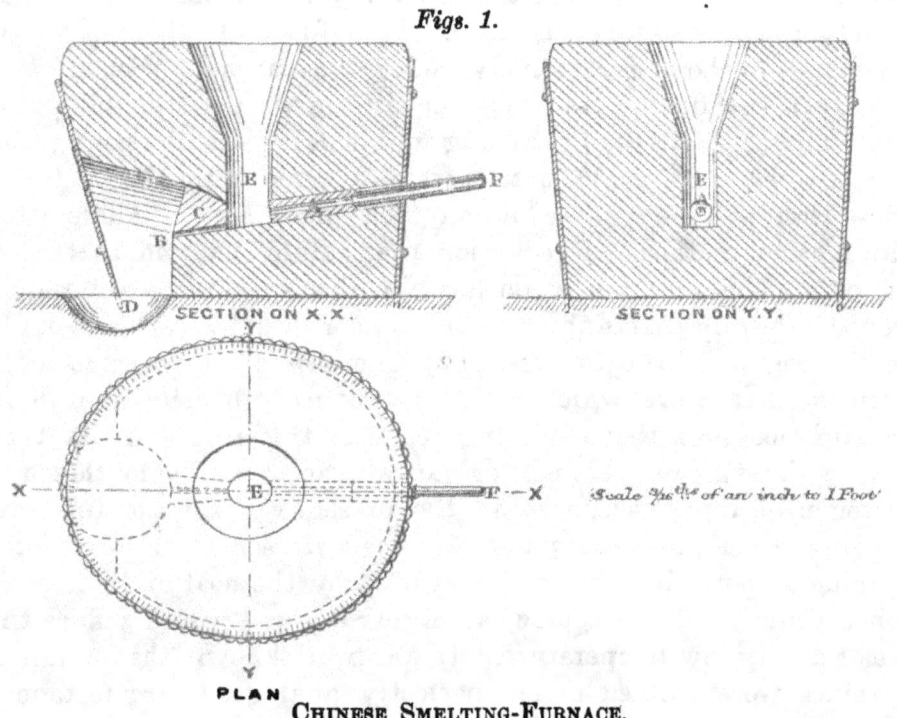

PLAN

CHINESE SMELTING-FURNACE.

with a wooden piston packed with leaves or feathers. One man mostly did the blowing, more rarely two. The blast was irregular and intermittent, and the average speed probably did not exceed ten strokes per minute. The front of the hearth was arched, and the crucible itself was closed in front, when at work, by a lump of clay, C, through which a small tap-hole, B, about ¾ inch in diameter, was kept open by means of a stick, or at times an iron rod. The tin trickled into a hole, D, in the ground lined with clay—the Chinese equivalent of the 'flote.' The molten tin was kept covered with burning charcoal, and from time to time was ladled out and cast into pigs by means of a sand mould, a wooden

Prof. Louis. block being used as a pattern.   Each pig weighed 60 kati (80 lbs.). The exact consumption of fuel was difficult to ascertain, and varied with the quality of the charcoal within wide limits.   He had been informed by some Chinamen that they used as little as 30 per cent., others as much as 100 per cent. of the weight of the ore.   Thus about 60 per cent. of tin was obtained from ore that probably contained 68 per cent. or 69 per cent., together with a small amount of very rich slag.   This slag was pounded under a rough tilt hammer, washed to extract the prills of metal, and then smelted in small furnaces about 2 feet 6 inches or 3 feet high, these poundings and smeltings being repeated between four and six times before the slag was thrown away as worthless.   Now, in watching an operation in one of these furnaces, the top would be found to be comparatively cold; the tap-hole was so cold that even the fusible iron and tin silicate were pasty, and would not run freely, all the heat being in a small reduction zone about the tuyeres.   He could not, therefore, agree with the Authors' view that a water-gas reduction took place here.   There were three other methods of reduction that might act, and that were more probable than the water-gas hypothesis :—(1) Direct reduction of the tin oxide by carbon or perhaps by carbonic-oxide in the region of the tuyeres.   (2) There was always some magnetite with the ore, which would be reduced to metallic iron in the furnace just above the tuyeres, and this would in its turn reduce the silicate of tin.   (3) It was most probable that the nitrogen of the atmosphere, in the presence of the alkaline carbonates in charcoal-ash, would combine with some carbon to form cyanide of potassium, which, volatilised by the heat of the tuyere zone, would condense somewhat higher up and would reduce the ore at a very low temperature.   It was well known that alkaline cyanides were formed under perfectly analogous circumstances in the blast-furnace, and the readiness with which such cyanides reduced oxide of tin was equally well known.   He had no doubt that all three of these reactions came into play in the Chinese method of tin-smelting, and had great doubts whether water-gas played any part at all in the reaction.   At the same time he might add that it seemed quite possible that reduction in a blast-furnace —perhaps with the injection of alkaline-cyanide to help the reaction—might ultimately prove a better method of tin-smelting than that of treatment in reverberatories.

Mr. Smith.   Mr. ERNEST A. SMITH observed that it had been stated by Messrs. McKillop and Ellis that "any ore containing arsenic or sulphur is thoroughly roasted at least once.   The furnace is of the 'blind

roaster' type, the ore being in a muffle, out of direct contact Mr. Smith.
with the fire." Although this form of furnace, from its gentle
and regular heat, no doubt possessed decided advantages in many
cases, yet its high cost of construction, and greater consumption
of fuel, were somewhat adverse to its employment. It would
be of interest, therefore, to know what special advantages were
derived from the use of the muffle furnace, for roasting the
ore under consideration, in place of the Oxland and Hocking
calciner, and other forms of roasting-furnaces in use in Cornwall
for the treatment of tin ores. The fact that the "cyanide"
method of assay was used exclusively for determining the value of
the ore was also of interest, as in Cornwall preference was usually
given to what was known as the "Cornish" method of assay,
although both methods were sometimes employed. The "button"
of tin obtained from ore assayed by the "cyanide" method was
practically pure when the assay was carefully conducted, and more
truly represented the percentage of tin present in the sample.
Yet it was claimed for the "Cornish" method that the resulting
tin "button" represented the quality of metal the smelter might
expect to obtain in the smelting of the ore, as this method of assay
was practically the "smelting operation" conducted on a very
small scale, powdered culm being used to effect the reduction of
the tin oxide. It would appear that the question of gaseous firing
merited a more extended trial.

Mr. CLEMES, in reply to the Correspondence, stated that the Mr. Clemes.
duty of the 1,000-lb. stamp-mill was between 2,500 lbs. and
3,000 lbs. per stamp in twenty-four hours. The same mill, with
wet crushing, would probably accomplish between 4,000 lbs. and
5,000 lbs. in twenty-four hours.

Mr. COURTNEY observed, in reply, that the controversy as to the Mr. Courtney.
nature of the deposits had not been referred to, as it would seem
to be never-ending. The latest information was undoubtedly
given by Professor Klockmann, who, after a visit to the mines, had
communicated his conclusions to the Berlin Academy of Science.
They were, however, opposed by others who had recently investi-
gated the subject, so that Mr. F. Roemer's enunciation of 1876 was
recalled, with the contrary opinion formed by Señor Gonzalo y
Tarin in 1887. He ventured the opinion that further investigation would tend to strengthen rather than otherwise the conclusions
of Señor Gonzalo y Tarin.

Mr. McKILLOP, in reply to the Correspondence, remarked that Mr. McKillop.
it had been considered unadvisable to enter into details on the
subject of the structure of the gas-furnace built at Pulo Brani.

Mr. McKillop. A furnace, however, could be built so that no metal could get into the chambers, and with experience the expenditure on repairs would be greatly reduced. The greater part of the expense at Pulo Brani was in connection with the producers, and by this time it was fair to assume that these were well enough understood to obviate any serious difficulty in adapting them to new conditions. The twelve months' experience at Pulo Brani convinced all concerned that the gas furnace was in principle entirely satisfactory; unfortunately the arrangement adopted in this particular case was really bad and it had to be condemned. It was undoubtedly true that some slags contained less than 5 per cent. of tin, but that amount was given as a real average. During the period in which the materials for the Paper had been collected, slags were obtained in which only mere traces of tin could be detected. But work of this high efficiency could not be expected regularly, especially with Javanese labour, and the Authors had no hesitation in expressing the opinion that an actual average of 5 per cent. would compare very favourably with the average in other works. Loss by volatilization appeared more serious than it really was. The flue was always closed during charging, and "tin fume," which rose from the surface of exposed metal, was not really a source of serious loss. In the first place, a watchful foreman could detect its formation at once and stop it by rabbling, and, in the second place, some experiments he had made, in conjunction with Mr. Ellis, showed clearly that an exposed surface of metal would apparently evolve large quantities of fume and still lose very slightly in weight. The question of running slags direct into secondary furnaces and treating them without allowing them to cool had been considered; it was thought to be perfectly feasible, but it demanded a very large scale of operations. The main cause of the backwardness of the metallurgy of tin was to be found in the smallness of the tin-works. A capable engineer in a large work would soon introduce these very obvious improvements. The remarks of Professor Louis were full of interest, and only emphasized the opinion expressed elsewhere that the metallurgical method by which by far the largest quantity of tin was produced was worthy of much more careful study than it had yet received. The muffle furnace had been adopted because it was simpler and less costly than a revolving calciner, for small quantities, and was much more easily controlled than an open-hearth roaster of the ordinary type.

LONDON: PRINTED BY WILLIAM CLOWES AND SONS, LIMITED, STAMFORD STREET
AND CHARING CROSS.

GOLD RUSH BOOKS

OREGON, USA

www.GoldMiningBooks.com

Books On Mining

Visit: www.goldminingbooks.com to order your copies or ask your favorite book seller to offer them.

Mining Books by Kerby Jackson

Gold Dust: Stories From Oregon's Mining Years - Oregon mining historian and prospector, Kerby Jackson, brings you a treasure trove of seventeen stories on Southern Oregon's rich history of gold prospecting, the prospectors and their discoveries, and the breathtaking areas they settled in and made homes. 5" X 8", 98 ppgs. Retail Price: $11.99

The Golden Trail: More Stories From Oregon's Mining Years - In his follow-up to "Gold Dust: Stories of Oregon's Mining Years", this time around, Jackson brings us twelve tales from Oregon's Gold Rush, including the story about the first gold strike on Canyon Creek in Grant County, about the old timers who found gold by the pail full at the Victor Mine near Galice, how Iradel Bray discovered a rich ledge of gold on the Coquille River during the height of the Rogue River War, a tale of two elderly miners on the hunt for a lost mine in the Cascade Mountains, details about the discovery of the famous Armstrong Nugget and others. 5" X 8", 70 ppgs. Retail Price: $10.99

Oregon Mining Books

Geology and Mineral Resources of Josephine County, Oregon - Unavailable since the 1970's, this important publication was originally compiled by the Oregon Department of Geology and Mineral Industries and includes important details on the economic geology and mineral resources of this important mining area in South Western Oregon. Included are notes on the history, geology and development of important mines, as well as insights into the mining of gold, copper, nickel, limestone, chromium and other minerals found in large quantities in Josephine County, Oregon. 8.5" X 11", 54 ppgs. Retail Price: $9.99

Mines and Prospects of the Mount Reuben Mining District - Unavailable since 1947, this important publication was originally compiled by geologist Elton Youngberg of the Oregon Department of Geology and Mineral Industries and includes detailed descriptions, histories and the geology of the Mount Reuben Mining District in Josephine County, Oregon. Included are notes on the history, geology, development and assay statistics, as well as underground maps of all the major mines and prospects in the vicinity of this much neglected mining district. 8.5" X 11", 48 ppgs. Retail Price: $9.99

The Granite Mining District - Notes on the history, geology and development of important mines in the well known Granite Mining District which is located in Grant County, Oregon. Some of the mines discussed include the Ajax, Blue Ribbon, Buffalo, Continental, Cougar-Independence, Magnolia, New York, Standard and the Tillicum. Also included are many rare maps pertaining to the mines in the area. 8.5" X 11", 48 ppgs. Retail Price: $9.99

Ore Deposits of the Takilma and Waldo Mining Districts of Josephine County, Oregon - The Waldo and Takilma mining districts are most notable for the fact that the earliest large scale mining of placer gold and copper in Oregon took place in these two areas. Included are details about some of the earliest large gold mines in the state such as the Llano de Oro, High Gravel, Cameron, Platerica, Deep Gravel and others, as well as copper mines such as the famous Queen of Bronze mine, the Waldo, Lily and Cowboy mines. This volume also includes six maps and 20 original illustrations. 8.5" X 11", 74 ppgs. Retail Price: $9.99

Metal Mines of Douglas, Coos and Curry Counties, Oregon - Oregon mining historian Kerby Jackson introduces us to a classic work on Oregon's mining history in this important re-issue of Bulletin 14C Volume 1, otherwise known as the Douglas, Coos & Curry Counties, Oregon Metal Mines Handbook. Unavailable since 1940, this important publication was originally compiled by the Oregon Department of Geology and Mineral Industries includes detailed descriptions, histories and the geology of over 250 metallic mineral mines and prospects in this rugged area of South West Oregon. 8.5" X 11", 158 ppgs. Retail Price: $19.99

Metal Mines of Jackson County, Oregon - Unavailable since 1943, this important publication was originally compiled by the Oregon Department of Geology and Mineral Industries includes detailed descriptions, histories and the geology of over 450 metallic mineral mines and prospects in Jackson County, Oregon. Included are such famous gold mining areas as Gold Hill, Jacksonville, Sterling and the Upper Applegate. 8.5" X 11", 220 ppgs. **Retail Price: $24.99**

Metal Mines of Josephine County, Oregon - Oregon mining historian Kerby Jackson introduces us to a classic work on Oregon's mining history in this important re-issue of Bulletin 14C, otherwise known as the Josephine County, Oregon Metal Mines Handbook. Unavailable since 1952, this important publication was originally compiled by the Oregon Department of Geology and Mineral Industries includes detailed descriptions, histories and the geology of over 500 metallic mineral mines and prospects in Josephine County, Oregon. 8.5" X 11", 250 ppgs. **Retail Price: $24.99**

Metal Mines of North East Oregon - Oregon mining historian Kerby Jackson introduces us to a classic work on Oregon's mining history in this important re-issue of Bulletin 14A and 14B, otherwise known as the North East Oregon Metal Mines Handbook. Unavailable since 1941, this important publication was originally compiled by the Oregon Department of Geology and Mineral Industries and includes detailed descriptions, histories and the geology of over 750 metallic mineral mines and prospects in North Eastern Oregon. 8.5" X 11", 310 ppgs. **Retail Price: $29.99**

Metal Mines of North West Oregon - Oregon mining historian Kerby Jackson introduces us to a classic work on Oregon's mining history in this important re-issue of Bulletin 14D, otherwise known as the North West Oregon Metal Mines Handbook. Unavailable since 1951, this important publication was originally compiled by the Oregon Department of Geology and Mineral Industries and includes detailed descriptions, histories and the geology of over 250 metallic mineral mines and prospects in North Western Oregon. 8.5" X 11", 182 ppgs. **Retail Price: $19.99**

Mines and Prospects of Oregon - Mining historian Kerby Jackson introduces us to a classic mining work by the Oregon Bureau of Mines in this important re-issue of The Handbook of Mines and Prospects of Oregon. Unavailable since 1916, this publication includes important insights into hundreds of gold, silver, copper, coal, limestone and other mines that operated in the State of Oregon around the turn of the 19th Century. Included are not only geological details on early mines throughout Oregon, but also insights into their history, production, locations and in some cases, also included are rare maps of their underground workings. 8.5" X 11", 314 ppgs. **Retail Price: $24.99**

Lode Gold of the Klamath Mountains of Northern California and South West Oregon (See California Mining Books)

Mineral Resources of South West Oregon - Unavailable since 1914, this publication includes important insights into dozens of mines that once operated in South West Oregon, including the famous gold fields of Josephine and Jackson Counties, as well as the Coal Mines of Coos County. Included are not only geological details on early mines throughout South West Oregon, but also insights into their history, production and locations. 8.5" X 11", 154 ppgs. **Retail Price: $11.99**

Chromite Mining in The Klamath Mountains of California and Oregon (See California Mining Books)

Southern Oregon Mineral Wealth - Unavailable since 1904, this rare publication provides a unique snapshot into the mines that were operating in the area at the time. Included are not only geological details on early mines throughout South West Oregon, but also insights into their history, production and locations. Some of the mining areas include Grave Creek, Greenback, Wolf Creek, Jump Off Joe Creek, Granite Hill, Galice, Mount Reuben, Gold Hill, Galls Creek, Kane Creek, Sardine Creek, Birdseye Creek, Evans Creek, Foots Creek, Jacksonville, Ashland, the Applegate River, Waldo, Kerby and the Illinois River, Althouse and Sucker Creek, as well as insights into local copper mining and other topics. 8.5" X 11", 64 ppgs. **Retail Price: $8.99**

Geology and Ore Deposits of the Takilma and Waldo Mining Districts - Unavailable since the 1933, this publication was originally compiled by the United States Geological Survey and includes details on gold and copper mining in the Takilma and Waldo Districts of Josephine County, Oregon. The Waldo and Takilma mining districts are most notable for the fact that the earliest large scale mining of placer gold and copper in Oregon took place in these two areas. Included in this report are details about some of the earliest large gold mines in the state such as the Llano de Oro, High Gravel, Cameron, Platerica, Deep Gravel and others, as well as copper mines such as the famous Queen of Bronze mine, the Waldo, Lily and Cowboy mines. In addition to geological examinations, insights are also provided into the production, day to day operations and early histories of these mines, as well as calculations of known mineral reserves in the area. This volume also includes six maps and 20 original illustrations. 8.5" X 11", 74 ppgs. **Retail Price: $9.99**

Gold Mines of Oregon - Oregon mining historian Kerby Jackson introduces us to a classic work on Oregon's mining history in this important re-issue of Bulletin 61, otherwise known as "Gold and Silver In Oregon". Unavailable since 1968, this important publication was originally compiled by geologists Howard C. Brooks and Len Ramp of the Oregon Department of Geology and Mineral Industries and includes detailed descriptions, histories and the geology of over 450 gold mines Oregon. Included are notes on the history, geology and gold production statistics of all the major mining areas in Oregon including the Klamath Mountains, the Blue Mountains and the North Cascades. While gold is where you find it, as every miner knows, the path to success is to prospect for gold where it was previously found. **8.5″ X 11″, 344 ppgs. Retail Price: $24.99**

Mines and Mineral Resources of Curry County Oregon - Originally published in 1916, this important publication on Oregon Mining has not been available for nearly a century. Included are rare insights into the history, production and locations of dozens of gold mines in Curry County, Oregon, as well as detailed information on important Oregon mining districts in that area such as those at Agness, Bald Face Creek, Mule Creek, Boulder Creek, China Diggings, Collier Creek, Elk River, Gold Beach, Rock Creek, Sixes River and elsewhere. Particular attention is especially paid to the famous beach gold deposits of this portion of the Oregon Coast. **8.5″ X 11″, 140 ppgs. Retail Price: $11.99**

Chromite Mining in South West Oregon - Originally published in 1961, this important publication on Oregon Mining has not been available for nearly a century. Included are rare insights into the history, production and locations of nearly 300 chromite mines in South Western Oregon. **8.5″ X 11″, 184 ppgs. Retail Price: $14.99**

Mineral Resources of Douglas County Oregon - Originally published in 1972, this important publication on Oregon Mining has not been available for nearly forty years. Included are rare insights into the geology, history, production and locations of numerous gold mines and other mining properties in Douglas County, Oregon. **8.5″ X 11″, 124 ppgs. Retail Price: $11.99**

Mineral Resources of Coos County Oregon - Originally published in 1972, this important publication on Oregon Mining has not been available for nearly forty years. Included are rare insights into the geology, history, production and locations of numerous gold mines and other mining properties in Coos County, Oregon. **8.5″ X 11″, 100 ppgs. Retail Price: $11.99**

Mineral Resources of Lane County Oregon - Originally published in 1938, this important publication on Oregon Mining has not been available for nearly seventy five years. Included are extremely rare insights into the geology and mines of Lane County, Oregon, in particular in the Bohemia, Blue River, Oakridge, Black Butte and Winberry Mining Districts. **8.5″ X 11″, 82 ppgs. Retail Price: $9.99**

Mineral Resources of the Upper Chetco River of Oregon: Including the Kalmiopsis Wilderness - Originally published in 1975, this important publication on Oregon Mining has not been available for nearly forty years. Withdrawn under the 1872 Mining Act since 1984, real insight into the minerals resources and mines of the Upper Chetco River has long been unavailable due to the remoteness of the area. Despite this, the decades of battle between property owners and environmental extremists over the last private mining inholding in the area has continued to pique the interest of those interested in mining and other forms of natural resource use. Gold mining began in the area in the 1850's and has a rich history in this geographic area, even if the facts surrounding it are little known. Included are twenty two rare photographs, as well as insights into the Becca and Morning Mine, the Emmly Mine (also known as Emily Camp), the Frazier Mine, the Golden Dream or Higgins Mine, Hustis Mine, Peck Mine and others. **8.5″ X 11″, 64 ppgs. Retail Price: $8.99**

Gold Dredging in Oregon - Originally published in 1939, this important publication on Oregon Mining has not been available for nearly seventy five years. Included are extremely rare insights into the history and day to day operations of the dragline and bucketline gold dredges that once worked the placer gold fields of South West and North East Oregon in decades gone by. Also included are details into the areas that were worked by gold dredges in Josephine, Jackson, Baker and Grant counties, as well as the economic factors that impacted this mining method. This volume also offers a unique look into the values of river bottom land in relation to both farming and mining, in how farm lands were mined, re-soiled and reclamated after the dredges worked them. Featured are hard to find maps of the gold dredge fields, as well as rare photographs from a bygone era. **8.5″ X 11″, 86 ppgs. Retail Price: $8.99**

Quick Silver Mining in Oregon - Originally published in 1963, this important publication on Oregon Mining has not been available for over fifty years. This publication includes details into the history and production of Elemental Mercury or Quicksilver in the State of Oregon. **8.5″ X 11″, 238 ppgs. Retail Price: $15.99**

Mines of the Greenhorn Mining District of Grant County Oregon - Originally published in 1948, this important publication on Oregon Mining has not been available for over sixty five years. In this publication are rare insights into the mines of the famous Greenhorn Mining District of Grant County, Oregon, especially the famous Morning Mine. Also included are details on the Tempest, Tiger, Bi-Metallic, Windsor, Psyche, Big Johnny, Snow Creek, Banzette and Paramount Mines, as well as prospects in the vicinities in the famous mining areas of Mormon Basin, Vinegar Basin and Desolation Creek. Included are hard to find mine maps and dozens of rare photographs from the bygone era of Grant County's rich mining history. **8.5″ X 11″, 72 ppgs. Retail Price: $9.99**

Geology of the Wallowa Mountains of Oregon: Part I (Volume 1) - Originally published in 1938, this important publication on Oregon Mining has not been available for nearly seventy five years. Included are details on the geology of this unique portion of North Eastern Oregon. This is the first part of a two book series on the area. Accompanying the text are rare photographs and historic maps. 8.5" X 11", 92 ppgs. **Retail Price: $9.99**

Geology of the Wallowa Mountains of Oregon: Part II (Volume 2) - Originally published in 1938, this important publication on Oregon Mining has not been available for nearly seventy five years. Included are details on the geology of this unique portion of North Eastern Oregon. This is the first part of a two book series on the area. Accompanying the text are rare photographs and historic maps. 8.5" X 11", 94 ppgs. **Retail Price: $9.99**

Field Identification of Minerals For Oregon Prospectors - Originally published in 1940, this important publication on Oregon Mining has not been available for nearly seventy five years. Included in this volume is an easy system for testing and identifying a wide range of minerals that might be found by prospectors, geologists and rockhounds in the State of Oregon, as well as in other locales. Topics include how to put together your own field testing kit and how to conduct rudimentary tests in the field. This volume is written in a clear and concise way to make it useful even for beginners. 8.5" X 11", 158 ppgs. **Retail Price: $14.99**

The Bohemia Mining District of Oregon - Originally published in 1900, this important publication on Oregon Mining has not been available for over a century. Included in this volume are important insights into the famous Bohemia Mining District of Oregon, including the histories and locations of important gold mines in the area such as the Ophir Mine, Clarence, Acturas, Peek-a-boo, White Swan, Combination Mine, the Musick Mine, The California, White Ghost, The Mystery, Wall Street, Vesuvius, Story, Lizzie Bullock, Delta, Elsie Dora, Golden Slipper, Broadway, Champion Mine, Knott, Noonday, Helena, White Wings, Riverside and others. Also included are notes on the nearby Blue River Mining District. 8.5" X 11", 58 ppgs. **Retail Price: $9.99**

The Gold Fields of Eastern Oregon - Unavailable since 1900, this publication was originally compiled by the Baker City Chamber of Commerce Offering important insights into the gold mining history of Eastern Oregon, "The Gold Fields of Eastern Oregon" sheds a rare light on many of the gold mines that were operating at the turn of the 19th Century in Baker County and Grant County in North Eastern Oregon. Some of the areas featured include the Cable Cove District, Baisely-Elhorn, Granite, Red Boy, Bonanza, Susanville, Sparta, Virtue, Vaughn, Sumpter, Burnt River, Rye Valley and other mining districts. Included is basic information on not only many gold mines that are well known to those interested in Eastern Oregon mining history, but also many mines and prospects which have been mostly lost to the passage of time. Accompanying are numerous rare photos 8.5" X 11", 78 ppgs. **Retail Price: $10.99**

Gold Mining in Eastern Oregon - Originally published in 1938, this important publication on Oregon Mining has not been available for over a century. Included in this volume are important insights into the famous mining districts of Eastern Oregon during the late 1930's. Particular attention is given to those gold mines with milling and concentrating facilities in the Greenhorn, Red Boy, Alamo, Bonanza, Granite, Cable Cove, Cracker Creek, Virtue, Keating, Medical Springs, Sanger, Sparta, Chicken Creek, Mormon Basin, Connor Creek, Cornucopia and the Bull Run Mining Districts. Some of the mines featured include the Ben Harrison, North Pole-Columbia, Highland Maxwell, Baisley-Elkhorn, White Swan, Balm Creek, Twin Baby, Gem of Sparta, New Deal, Gleason, Gifford-Johnson, Cornucopia, Record, Bull Run, Orion and others. Of particular interest are the mill flow sheets and descriptions of milling operations of these mines. 8.5" X 11", 68 ppgs. **Retail Price: $8.99**

The Gold Belt of the Blue Mountains of Oregon - Originally published in 1901, this important publication on Oregon Mining has not been available for over a century. Included in this volume are rare insights into the gold deposits of the Blue Mountains of North East Oregon, including the history of their early discovery and early production. Extensive details are offered on this important mining area's mineralogy and economic geology, as well as insights into nearby gold placers, silver deposits and copper deposits. Featured are the Elkhorn and Rock Creek mining districts, the Pocahontas district, Auburn and Minersville districts, Sumpter and Cracker Creek, Cable Cove, the Camp Carson district, Granite, Alamo, Greenhorn, Robinsonville, the Upper Burnt River Valley and Bonanza districts, Susanville, Quartzburg, Canyon Creek, Virtue, the Copper Butte district, the North Powder River, Sparta, Eagle Creek, Cornucopia, Pine Creek, Lower Powder River, the Upper Snake River Canyon, Rye Valley, Lower Burnt River Valley, Mormon Basin, the Malheur and Clarks Creek districts, Sutton Creek and others. Of particular interest are important details on numerous gold mines and prospects in these mining districts, including their locations, histories, geology and other important information, as well as information on silver, copper and fire opal deposits. 8.5" X 11", 250 ppgs. **Retail Price: $24.99**

Mining in the Cascades Range of Oregon - Originally published in 1938, this important publication on Oregon Mining has not been available for over seventy five years. Included in this volume are rare insights into the gold mines and other types of metal mines in the Cascades Mountain Range of Oregon. Some of the important mining areas covered include the famous Bohemia Mining District, the North Santiam Mining District, Quartzville Mining District, Blue River Mining District, Fall Creek Mining District, Oakridge District, Zinc District, Buzzard-Al Sarena District, Grand Cove, Climax District and Barron Mining District. Of particular interest are important details on over 100 mines and prospects in these mining districts, including their locations, histories, geology and other important information. **8.5" X 11", 170 ppgs. Retail Price: $14.99**

Beach Gold Placers of the Oregon Coast - Originally published in 1934, this important publication on Oregon Mining has not been available for over 80 years. Included in this volume are rare insights into the beach gold deposits of the State of Oregon, including their locations, occurance, composition and geology. Of particular interest is information on placer platinum in Oregon's rich beach deposits. Also included are the locations and other information on some famous Oregon beach mines, including the Pioneer, Eagle, Chickamin, Iowa and beach placer mines north of the mouth of the Rogue River. **8.5" X 11", 60 ppgs. Retail Price: $8.99**

Idaho Mining Books

Gold in Idaho - Unavailable since the 1940's, this publication was originally compiled by the Idaho Bureau of Mines and includes details on gold mining in Idaho. Included is not only raw data on gold production in Idaho, but also valuable insight into where gold may be found in Idaho, as well as practical information on the gold bearing rocks and other geological features that will assist those looking for placer and lode gold in the State of Idaho. This volume also includes thirteen gold maps that greatly enhance the practical usability of the information contained in this small book detailing where to find gold in Idaho. **8.5" X 11", 72 ppgs. Retail Price: $9.99**

Geology of the Couer D'Alene Mining District of Idaho - Unavailable since 1961, this publication was originally compiled by the Idaho Bureau of Mines and Geology and includes details on the mining of gold, silver and other minerals in the famous Coeur D'Alene Mining District in Northern Idaho. Included are details on the early history of the Coeur D'Alene Mining District, local tectonic settings, ore deposit features, information on the mineral belts of the Osburn Fault, as well as detailed information on the famous Bunker Hill Mine, the Dayrock Mine, Galena Mine, Lucky Friday Mine and the infamous Sunshine Mine. This volume also includes sixteen hard to find maps. **8.5" X 11", 70 ppgs. Retail Price: $9.99**

The Gold Camps and Silver Cities of Idaho - Originally published in 1963, this important publication on Idaho Mining has not been available for nearly fifty years. Included are rare insights into the history of Idaho's Gold Rush, as well as the mad craze for silver in the Idaho Panhandle. Documented in fine detail are the early mining excitements at Boise Basin, at South Boise, in the Owyhees, at Deadwood, Long Valley, Stanley Basin and Robinson Bar, at Atlanta, on the famous Boise River, Volcano, Little Smokey, Banner, Boise Ridge, Hailey, Leesburg, Lemhi, Pearl, at South Mountain, Shoup and Ulysses, Yellow Jacket and Loon Creek. The story follows with the appearance of Chinese miners at the new mining camps on the Snake River, Black Pine, Yankee Fork, Bay Horse, Clayton, Heath, Seven Devils, Gibbonsville, Vienna and Sawtooth City. Also included are special sections on the Idaho Lead and Silver mines of the late 1800's, as well as the mining discoveries of the early 1900's that paved the way for Idaho's modern mining and mineral industry. Lavishly illustrated with rare historic photos, this volume provides a one of a kind documentary into Idaho's mining history that is sure to be enjoyed by not only modern miners and prospectors who still scour the hills in search of nature's treasures, but also those enjoy history and tromping through overgrown ghost towns and long abandoned mining camps. **8.5" X 11", 186 ppgs. Retail Price: $14.99**

Ore Deposits and Mining in North Western Custer County Idaho - Unavailable since 1913, this important publication was originally published by the Us Department of the Interior and has been unavailable for a century. Included are fine details on the geology, geography, gold placers and gold and silver bearing quartz veins of the mining region of North West Custer County, Idaho. Of particular interest is a rare look at the mines and prospects of the region, including those such as the Ramshorn Mine, SkyLark, Riverview, Excelsior, Beardsley, Pacific, Hoosier, Silver Brick, Forest Rose and dozens of others in the Bay Horse Mining District. Also covered are the mines of the Yankee Fork District such as the Lucky Boy, Badger, Black, Enterprise, Charles Dickens, Morrison, Golden Sunbeam, Montana, Golden Gate and others, as well as those in the Loon Mining District. **8.5" X 11", 126 ppgs. Retail Price: $12.99**

Gold Rush To Idaho - Unavailable since 1963, this important publication was originally published by the Idaho Bureau of Mines and has been unavailable for 50 years. "Gold Rush To Idaho" revisits the earliest years of the discovery of gold in Idaho Territory and introduces us to the conditions that the pioneer gold seekers met when they blazed a trail through the wilderness of Idaho's mountains and discovered the precious yellow metal at Oro Fino and Pierce. Subsequent rushes followed at places like Elk City, Newsome, Clearwater Station, Florence, Warrens and elsewhere. Of particular interest is a rare look at the hardships that the first miners in Idaho met with during their day to day existences and their attempts to bring law and order to their mining camps. **8.5" X 11", 88 ppgs. Retail Price: $9.99**

The Geology and Mines of Northern Idaho and North Western Montana - Unavailable since 1909, this important publication was originally published by the Us Department of the Interior and has been unavailable for a century. Included are fine details on the geology and geography of the mining regions of Northern Idaho and North Western Montana. Of particular interest is a rare look at the mines and prospects of the region, including those in the Pine Creek Mining District, Lake Pend Oreille district, Troy Mining District, Sylvanite District, Cabinet Mining District, Prospect Mining District and the Missoula Valley. Some of the mines featured include the Iron Mountain, Silver Butte, Snowshoe, Grouse Mountain Mine and others. **8.5" X 11", 142 ppgs. Retail Price: $12.99**

Mining in the Alturas Quadrangle of Blaine County Idaho - Unavailable since 1922, this important publication was originally published by the Idaho Bureau of Mines and has been unavailable for ninety years. Topics include the geology, rock formations and the formation of ore deposits in this important mining area of Idaho. Of particular focus is information on the local geology, quartz veins and ore deposits of this portion of Idaho. Included are hard to find details, including the descriptions and locations of numerous gold and silver mines in the area including the Silver King, Pilgrim, Columbia, Lone Jack, Sunbeam, Pride of the West, Lucky Boy, Scotia, Atlanta, Beaver-Bidwell and others mines and prospects. **8.5" X 11", 56 ppgs. Retail Price: $8.99**

Mining in Lemhi County Idaho - Originally published in 1913, this important book on Idaho Mining has not been available to miners for over a century. Included are rare insights into hundreds of gold, silver, copper and other mines in this famous Idaho mining area. Details include the locations, geology, history, production and other facts of the mines of this region, not only gold and silver hardrock mines, but also gold placer mines, lead-silver deposits, copper mines, cobalt-nickel deposits, tungsten and tin mines . It is lavishly illustrated with hard to find photos of the period and rare mining maps. Some of the vicinities featured include the Nicholia Mining District, Spring Mountain District, Texas District, Blue Wing District, Junction District, McDevitt District, Pratt Creek, Eldorado District, Kirtley Creek, Carmen Creek, Gibbonsville, Indian Creek, Mineral Hill District, Mackinaw, Eureka District, Blackbird District, YellowJacket District, Gravel Range District, Junction District, Parker Mountain and other mining districts. **8.5" X 11", 226 ppgs. Retail Price: $19.99**

Utah Mining Books

Fluorite in Utah - Unavailable since 1954, this publication was originally compiled by the USGS, State of Utah and U.S. Atomic Energy Commission and details the mining of fluorspar, also known as fluorite in the State of Utah. Included are details on the geology and history of fluorspar (fluorite) mining in Utah, including details on where this unique gem mineral may be found in the State of Utah. **8.5" X 11", 60 ppgs. Retail Price: $8.99**

California Mining Books

The Tertiary Gravels of the Sierra Nevada of California - Mining historian Kerby Jackson introduces us to a classic mining work by Waldemar Lindgren in this important re-issue of The Tertiary Gravels of the Sierra Nevada of California. Unavailable since 1911, this publication includes details on the gold bearing ancient river channels of the famous Sierra Nevada region of California. **8.5" X 11", 282 ppgs. Retail Price: $19.99**

The Mother Lode Mining Region of California - Unavailable since 1900, this publication includes details on the gold mines of California's famous Mother Lode gold mining area. Included are details on the geology, history and important gold mines of the region, as well as insights into historic mining methods, mine timbering, mining machinery, mining bell signals and other details on how these mines operated. Also included are insights into the gold mines of the California Mother Lode that were in operation during the first sixty years of California's mining history. **8.5" X 11", 176 ppgs. Retail Price: $14.99**

Lode Gold of the Klamath Mountains of Northern California and South West Oregon - Unavailable since 1971, this publication was originally compiled by Preston E. Hotz and includes details on the lode mining districts of Oregon and California's Klamath Mountains. Included are details on the geology, history and important lode mines of the French Gulch, Deadwood, Whiskeytown, Shasta, Redding, Muletown, South Fork, Old Diggings, Dog Creek (Delta), Bully Choop (Indian Creek), Harrison Gulch, Hayfork, Minersville, Trinity Center, Canyon Creek, East Fork, New River, Denny, Liberty (Black Bear), Cecilville, Callahan, Yreka, Fort Jones and Happy Camp mining districts in California, as well as the Ashland, Rogue River, Applegate, Illinois River, Takilma, Greenback, Galice, Silver Peak, Myrtle Creek and Mule Creek districts of South Western Oregon. Also included are insights into the mineralization and other characteristics of this important mining region. **8.5" X 11", 100 ppgs. Retail Price: $10.99**

Mines and Mineral Resources of Shasta County, Siskiyou County, Trinity County: California - Unavailable since 1915, this publication was originally compiled by the California State Mining Bureau and includes details on the gold mines of this area of Northern California. Also included are insights into the mineralization and other characteristics of this important mining region, as well as the location of historic gold mines. **8.5" X 11", 204 ppgs. Retail Price: $19.99**

Geology of the Yreka Quadrangle, Siskiyou County, California - Unavailable since 1977, this publication was originally compiled by Preston E. Hotz and includes details on the geology of the Yreka Quadrangle of Siskiyou County, California. Also included are insights into the mineralization and other characteristics of this important mining region. **8.5" X 11", 78 ppgs. Retail Price: $7.99**

Mines of San Diego and Imperial Counties, California - Originally published in 1914, this important publication on California Mining has not been available for a century. This publication includes important information on the early gold mines of San Diego and Imperial County, which were some of the first gold fields mined in California by early Spanish and Mexican miners before the 49ers came on the scene. Included are not only details on early mining methods in the area, production statistics and geological information, but also the location of the early gold mines that helped make California "The Golden State". Also included are details on the mining of other minerals such as silver, lead, zinc, manganese, tungsten, vanadium, asbestos, barite, borax, cement, clay, dolomite, fluospar, gem stones, graphite, marble, salines, petroleum, stronium, talc and others. **8.5" X 11", 116 ppgs. Retail Price: $12.99**

Mines of Sierra County, California - Unavailable since 1920, this publication was originally compiled by the California State Mining Bureau and includes details on the gold mines of Sierra County, California. Also included are insights into the mineralization and other characteristics of this important mining region, as well as the location of historic gold mines. **8.5" X 11", 156 ppgs. Retail Price: $19.99**

Mines of Plumas County, California - Unavailable since 1918, this publication was originally compiled by the California State Mining Bureau and includes details on the gold mines of Plumas County, California. Also included are insights into the mineralization and other characteristics of this important mining region, as well as the location of historic gold mines. **8.5" X 11", 200 ppgs. Retail Price: $19.99**

Mines of El Dorado, Placer, Sacramento and Yuba Counties, California - Originally published in 1917, this important publication on California Mining has not been available for nearly a century. This publication includes important information on the early gold mines of El Dorado County, Placer County, Sacramento County and Yuba County, which were some of the first gold fields mined by the Forty-Niners during the California Gold Rush. Included are not only details on early mining methods in the area, production statistics and geological information, but also the location of the early gold mines that helped make California "The Golden State". Also included are insights into the early mining of chrome, copper and other minerals in this important mining area. **8.5" X 11", 204 ppgs. Retail Price: $19.99**

Mines of Los Angeles, Orange and Riverside Counties, California - Originally published in 1917, this important publication on California Mining has not been available for nearly a century. This publication includes important information on the early gold mines of Los Angeles County, Orange County and Riverside County, which were some of the first gold fields mined in California by early Spanish and Mexican miners before the 49ers came on the scene. Included are not only details on early mining methods in the area, production statistics and geological information, but also the location of the early gold mines that helped make California "The Golden State". **8.5" X 11", 146 ppgs. Retail Price: $12.99**

Mines of San Bernadino and Tulare Counties, California - Originally published in 1917, this important publication on California Mining has not been available for nearly a century. This publication includes important information on the early gold mines of San Bernadino and Tulare County, which were some of the first gold fields mined in California by early Spanish and Mexican miners before the 49ers came on the scene. Included are not only details on early mining methods in the area, production statistics and geological information, but also the location of the early gold mines that helped make California "The Golden State". Also included are details on the mining of other minerals such as copper, iron, lead, zinc, manganese, tungsten, vanadium, asbestos, barite, borax, cement, clay, dolomite, fluospar, gem stones, graphite, marble, salines, petroleum, stronium, talc and others. **8.5" X 11", 200 ppgs. Retail Price: $19.99**

Chromite Mining in The Klamath Mountains of California and Oregon - Unavailable since 1919, this publication was originally compiled by J.S. Diller of the United States Department of Geological Survey and includes details on the chromite mines of this area of Northern California and Southern Oregon. Also included are insights into the mineralization and other characteristics of this important mining region, as well as the location of historic mines. Also included are insights into chromite mining in Eastern Oregon and Montana. **8.5" X 11", 98 ppgs. Retail Price: $9.99**

Mines and Mining in Amador, Calaveras and Tuolumne Counties, California - Unavailable since 1915, this publication was originally compiled by William Tucker and includes details on the mines and mineral resources of this important California mining area. Included are details on the geology, history and important gold mines of the region, as well as insights into other local mineral resources such as asbestos, clay, copper, talc, limestone and others. Also included are insights into the mineralization and other characteristics of this important portion of California's Mother Lode mining region. 8.5" X 11", 198 ppgs. Retail Price: $14.99

The Cerro Gordo Mining District of Inyo County California - Unavailable since 1963, this publication was originally compiled by the United States Department of Interior. Included are insights into the mineralization and other characteristics of this important mining region of Southern California. Topics include the mining of gold and silver in this important mining district in Inyo County, California, including details on the history, production and locations of the Cerro Gordo Mine, the Morning Star Mine, Estelle Tunnel, Charles Lease Tunnel, Ignacio, Hart, Crosscut Tunnel, Sunset, Upper Newtown, Newtown, Ella, Perseverance, Newsboy, Belmont and other silver and gold mines in the Cerro Gordo Mining District. This volume also includes important insights into the fossil record, geologic formations, faults and other aspects of economic geology in this California mining district. 8.5" X 11", 104 ppgs. Retail Price: $10.99

Mining in Butte, Lassen, Modoc, Sutter and Tehama Counties of California - Unavailable since 1917, this publication was originally compiled by the United States Department of Interior. Included are insights into the mineralization and other characteristics of this important mining region of California. Topics include the mining of asbestos, chromite, gold, diamonds and manganese in Butte County, the mining of gold and copper in the Hayden Hill and Diamond Mountain mining districts of Lassen County, the mining of coal, salt, copper and gold in the High Grade and Winters mining districts of Modoc County, gold mining in Sutter County and the mining of gold, chromite, manganese and copper in Tehama County. This volume also includes the production records and locations of numerous mines in this important mining region. 8.5" X 11", 114 ppgs. Retail Price: $11.99

Mines of Trinity County California - Originally published in 1965, this important publication on California Mining has not been available for nearly fifty years. This publication includes important information on mines and mining in Trinity County, California, as well insights into the mineralization and geology of this important mining area in Northern California. Included are extensive details on hardrock and placer gold mines and prospects, including charts showing the locations of these historic mines.. 8.5" X 11", 144 ppgs. Retail Price: $12.99

Mines of Kern County California - Originally published in 1962, this important publication on California Mining has not been available for nearly fifty years. This publication includes important information on mines and mining in Kern County, California, as well insights into the mineralization and geology of this important mining area in California. Included are extensive details on hardrock and placer gold mines and prospects, including charts showing the locations of these historic mines. 8.5" X 11", 398 ppgs. Retail Price: $24.99

Mines of Calaveras County California - Originally published in 1962, this important publication on California Mining has not been available for nearly fifty years. This publication includes important information on mines and mining in Calaveras County, California, as well insights into the mineralization and geology of this important mining area in Northern California. Included are extensive details on hardrock and placer gold mines and prospects, including charts showing the locations of these historic mines. 8.5" X 11", 236 ppgs. Retail Price: $19.99

Lode Gold Mining in Grass Valley California - Unavailable since 1940, this publication was originally compiled by the United States Department of Interior. Included are insights into the gold mineralization and other characteristics of this important mining region of Nevada County, California. This volume also includes important insights into the geologic formations, faults and other aspects of economic geology in this California mining district. Of particular interest are the fine details on many hardrock gold mines in the area, including their locations, histories, development and mineralization. Some of the mines featured include the Gold Hill Mine, Massachusetts Hill, Boundary, Peabody, Golden Center, North Star, Omaha, Lone Jack, Homeward Bound, Hartery, Wisconsin, Allison Ranch, Phoenix, Kate Hayes, W.Y.O.D., Empire, Rich Hill, Daisy Hill, Orleans, Sultana, Centennial, Conlin, Ben Franklin, Crown Point and many others. 8.5" X 11", 148 ppgs. Retail Price: $12.99

Lode Mining in the Alleghany District of Sierra County California - Unavailable since 1913, this publication was originally compiled by the United States Department of Interior. Included are insights into the mineralization and other characteristics of this important mining region of Sierra County. Included are details on the history, production and locations of numerous hardrock gold mines in this famous California area, including the Tightner Mine, Minnie D., Osceola, Eldorado, Twenty One, Sherman, Kenton, Oriental, Rainbow, Plumbago, Irelan, Gold Canyon, North Fork, Federal, Kate Hardy and others. This volume also includes important insights into the fossil record, geologic formations, faults and other aspects of economic geology in this California mining district. 8.5" X 11", 48 ppgs. Retail Price: $7.99

Six Months In The Gold Mines During The California Gold Rush - Unavailable since 1850, this important work is a first hand account of one "49'ers" personal experience during the great California Gold Rush, shedding important light on one of the most exciting periods in the history of not only California, but also the world. Compiled from journals written between 1847 and 1849 by E. Gould Buffum, a native of New York, "Six Months In The Gold Mines During The California Gold Rush" offers a rare look into the day to day lives of the people who came to California to work in her gold mines when the state was still a great frontier. **8.5" X 11", 290 ppgs. Retail Price: $19.99**

Quartz Mines of the Grass Valley Mining District of California - Unavailable since 1867, this important publication has not been available since those days. This rare publication offers a short dissertation on the early hardrock mines in this important mining district in the California Mother Lode region between the 1850's and 1860's. Also included are hard to find details on the mineralization and locations of these mines, as well as how they were operated in those day. **8.5" X 11", 44 ppgs. Retail Price: $8.99**

Alaska Mining Books

Ore Deposits of the Willow Creek Mining District, Alaska - Unavailable since 1954, this hard to find publication includes valuable insights into the Willow Creek Mining District near Hatcher Pass in Alaska. The publication includes insights into the history, geology and locations of the well known mines in the area, including the Gold Cord, Independence, Fern, Mabel, Lonesome, Snowbird, Schroff-O'Neil, High Grade, Marion Twin, Thorpe, Webfoot, Kelly-Willow, Lane, Holland and others. **8.5" X 11", 96 ppgs. Retail Price: $9.99**

The Juneau Gold Belt of Alaska - Unavailable since 1906, this hard to find publication includes valuable insights into the gold mines around Juneau, Alaska. The publication includes important details into the history, geology and locations of the well known gold mines and prospects in the area, including those around Windham Bay, Holkham Bay, Port Snettisham, on Grindstone and Rhine Creeks, Gold Creek, Douglas Island, Salmon Creek, Lemon Creek, Nugget Creek, from the Mendenhall River to Berners Bay, McGinnis Creek, Montana Creek, Peterson Creek, Windfall Creek, the Eagle River, Yankee Basin, Yankee Curve, Kowee Creek and elsewhere. Not only are gold placer mines included, but also hardrock gold mines. **8.5" X 11", 224 ppgs. Retail Price: $19.99**

Arizona Mining Books

Mines and Mining in Northern Yuma County Arizona - Originally published in 1911, this important publication on Arizona Mining has not been available for over a hundred years. Included are rare insights into the gold, silver, copper and quicksilver mines of Yuma County, Arizona together with hard to find maps and photographs. Some of the mines and mining districts featured include the Planet Copper Mine, Mineral Hill, the Clara Consolidated Mine, Viati Mine, Copper Basin prospect, Bowman Mine, Quartz King, Billy Mack, Carnation, the Wardwell and Osbourne, Valensuella Copper, the Mariquita, Colonial Mine, the French American, the New York-Plomosa, Guadalupe, Lead Camp, Mudersbach Copper Camp, Yellow Bird, the Arizona Northern (Salome Strike), Bonanza (Harqua Hala), Golden Eagle, Hercules, Socorro and others. **8.5" X 11", 144 ppgs. Retail Price: $11.99**

The Aravaipa and Stanley Mining Districts of Graham County Arizona - Originally published in 1925, this important publication on Arizona Mining has not been available for nearly ninety years. Included are rare insights into the gold and silver mines of these two important mining districts, together with hard to find maps. **8.5" X 11", 140 ppgs. Retail Price: $11.99**

Gold in the Gold Basin and Lost Basin Mining Districts of Mohave County, Arizona - This volume contains rare insights into the geology and gold mineralization of the Gold Basin and Lost Basin Mining Districts of Mohave County, Arizona that will be of benefit to miners and prospectors. Also included is a significant body of information on the gold mines and prospects of this portion of Arizona. This volume is lavishly illustrated with rare photos and mining maps. **8.5" X 11", 188 ppgs. Retail Price: $19.99**

Mines of the Jerome and Bradshaw Mountains of Arizona - This important publication on Arizona Mining has not been available for ninety years. This volume contains rare insights into the geology and ore deposits of the Jerome and Bradshaw Mountains of Arizona that will be of benefit to miners and prospectors who work those areas. Included is a significant body of information on the mines and prospects of the Verde, Black Hills, Cherry Creek, Prescott, Walker, Groom Creek, Hassayampa, Bigbug, Turkey Creek, Agua Fria, Black Canyon, Peck, Tiger, Pine Grove, Bradshaw, Tintop, Humbug and Castle Creek Mining Districts. This volume is lavishly illustrated with rare photos and mining maps. **8.5" X 11", 218 ppgs. Retail Price: $19.99**

The Ajo Mining District of Pima County Arizona - This important publication on Arizona Mining has not been available for nearly seventy years. This volume contains rare insights into the geology and mineralization of the Ajo Mining District in Pima County, Arizona and in particular the famous New Cornelia Mine. **8.5" X 11", 126 ppgs. Retail Price: $11.99**

Mining in the Santa Rita and Patagonia Mountains of Arizona - Originally published in 1915, this important publication on Arizona Mining has not been available for nearly a century. Included are rare insights into hundreds of gold, silver, copper and other mines in this famous Arizona mining area. Details include the locations, geology, history, production and other facts of the mines of this region. 8.5" X 11", 394 ppgs. Retail Price: $24.99

Mining in the Bisbee Quadrangle of Arizona - Originally published in 1906, this important publication on Arizona Mining has not been available for nearly a century. Included are rare insights into hundreds of gold, silver, copper and other mines in this famous Arizona mining area. Details include the locations, geology, history, production and other facts of the mines of this important mining region. 8.5" X 11", 188 ppgs. Retail Price: $14.99

Montana Mining Books

A History of Butte Montana: The World's Greatest Mining Camp - First published in 1900 by H.C. Freeman, this important publication sheds a bright light on one of the most important mining areas in the history of The West. Together with his insights, as well as rare photographs of the periods, Harry Freeman describes Butte and its vicinity from its early beginnings, right up to its flush years when copper flowed from its mines like a river. At the time of publication, Butte, Montana was known worldwide as "The Richest Mining Spot On Earth" and produced not only vast amounts of copper, but also silver, gold and other metals from its mines. Freeman illustrates, with great detail, the most important mines in the vicinity of Butte, providing rare details on their owners, their history and most importantly, how the mines operated and how their treasures were extracted. Of particular interest are the dozens of rare photographs that depict mines such as the famous Anaconda, the Silver Bow, the Smoke House, Moose, Paulin, Buffalo, Little Minah, the Mountain Consolidated, West Greyrock, Cora, the Green Mountain, Diamond, Bell, Parnell, the Neversweat, Nipper, Original and many others. 8.5" X 11", 142 ppgs. Retail Price: $12.99

The Butte Mining District of Montana - This important publication on Montana Mining has not been available for over a century. Included are rare insights into the gold, copper and silver mines of Butte, Montana together with hard to find maps and photographs. Some of the topics include the early history of gold, silver and copper mining in the Butte area, insight into the geology of its mining areas, the local distribution of gold, silver and copper ores, as well their composition and how to identify them. Also included are detailed facts about the mines in the Butte Mining District, including the famous Anaconda Mine, Gagnon, Parrot, Blue Vein, Moscow, Poulin, Stella, Buffalo, Green Mountain, Wake Up Jim, the Diamond-Bell Group, Mountain Consolidated, East Greyrock, West Greyrock, Snowball, Corra, Speculator, Adirondack, Miners Union, the Jessie-Edith May Group, Otisco, Iduna, Colorado, Lizzie, Cambers, Anderson, Hesperus, Preferencia and dozens of others. 8.5" X 11", 298 ppgs. Retail Price: $24.99

Mines of the Helena Mining Region of Montana - This important publication on Montana Mining has not been available for over a century. Included are rare insights into the gold, copper and silver mines of the vicinity of Helena, Montana, including the Marysville Mining District, Elliston Mining District, Rimini Mining District, Helena Mining District, Clancy Mining District, Wickes Mining District, Boulder and Basin Mining Districts and the Elkhorn Mining District. Some of the topics include the early history of gold, silver and copper mining in the Helena area, insight into the geology of its mining areas, the local distribution of gold, silver and copper ores, as well their composition and how to identify them. Also included are detailed facts, history, geology and locations of over one hundred gold, silver and copper mines in the area . 8.5" X 11", 162 ppgs, Retail Price: $14.99

Mines and Geology of the Garnet Range of Montana - This important publication on Montana Mining has not been available for over a century. Included are rare insights into the gold, copper and silver mines of the vicinity of this important mining area of Montana. Some of the topics include the early history of gold, silver and copper mining in the Garnet Mountains, insight into the geology of its mining areas, the local distribution of gold, silver and copper ores, as well their composition and how to identify them. Also included are detailed facts, history, geology and locations of numerous gold, silver and copper mines in the area . 8.5" X 11", 100 ppgs, Retail Price: $11.99

Mines and Geology of the Philipsburg Quadrangle of Montana - This important publication on Montana Mining has not been available for over a century. Included are rare insights into the gold, copper and silver mines of the vicinity of this important mining area of Montana. Some of the topics include the early history of gold, silver and copper mining in the Philipsburg Quadrangle, insight into the geology of its mining areas, the local distribution of gold, silver and copper ores, as well their composition and how to identify them. Also included are detailed facts, history, geology and locations of over one hundred gold, silver and copper mines in the area 8.5" X 11", 290 ppgs, Retail Price: $24.99

Geology of the Marysville Mining District of Montana - Included are rare insights into the mining geology of the Marysville Mining District. Some of the topics include the early history of gold, silver and copper mining in the area, insight into the geology of its mining areas, the local distribution of gold, silver and copper ores, as well their composition and how to identify them. Also included are detailed facts, history, geology and locations of gold, silver and copper mines in the area 8.5" X 11", 198 ppgs, Retail Price: $19.99

The Geology and Mines of Northern Idaho and North Western Montana

See listing under Idaho.

Nevada Mining Books

The Bull Frog Mining District of Nevada - Unavailable since 1910, this publication was originally compiled by the United States Department of Interior. This volume also includes important insights into the geologic formations, faults and other aspects of economic geology in this Nevada mining district. Of particular interest are the fine details on many mines in the area, including their locations, histories, development and mineralization. Some of the mines featured include the National Bank Mine, Providence, Gibraltor, Tramps, Denver, Original Bullfrog, Gold Bar, Mayflower, Homestake-King and other mines and prospects. **8.5" X 11", 152 ppgs, Retail Price: $14.99**

History of the Comstock Lode - Unavailable since 1876, this publication was originally released by John Wiley & Sons. This volume also includes important insights into the famous Comstock Lode of Nevada that represented the first major silver discovery in the United States. During its spectacular run, the Comstock produced over 192 million ounces of silver and 8.2 million ounces of gold. Not only did the Comstock result in one of the largest mining rushes in history and yield immense fortunes for its owners, but it made important contributions to the development of the State of Nevada, as well as neighboring California. Included here are important details on not only the early development and history of the Comstock, but also rare early insight into its mines, ore and its geology. **8.5" X 11", 244 ppgs, Retail Price: $19.99**

Colorado Mining Books

Ores of The Leadville Mining District - Unavailable since 1926, this publication was originally compiled by the United States Department of Interior. This volume also includes important insights into the ores and mineralization of the Leadville Mining District in Colorado. Topics include historic ore prospecting methods, local geology, insights into ore veins and stockworks, the local trend and distribution of ore channels, reverse faults, shattered rock above replacement ore bodies, mineral enrichment in oxidized and sulphide zones and more. **8.5" X 11", 66 ppgs, Retail Price: $8.99**

Mining in Colorado - Unavailable since 1926, this publication was originally compiled by the United States Department of Interior. This volume also includes important insights into the mining history of Colorado from its early beginnings in the 1850's right up to the mid 1920's. Not only is Colorado's gold mining heritage included, but also its silver, copper, lead and zinc mining industry. Each mining area is treated separately, detailing the development of Colorado's mines on a county by county basis. **8.5" X 11", 284 ppgs, Retail Price: $19.99**

Gold Mining in Gilpin County Colorado - Unavailable since 1876, this publication was originally compiled by the Register Steam Printing House of Central City, Colorado. A rare glimpse at the gold mining history and early mines of Gilpin County, Colorado from their first discovery in the 1850's up to the "flush years" of the mid 1870's. Of particular interest is the history of the discovery of gold in Gilpin County and details about the men who made those first strikes. Special focus is given to the early gold mines and first mining districts of the area, many of which are not detailed in other books on Colorado's gold mining history. **8.5" X 11", 156 ppgs, Retail Price: $12.99**

Mining in the Gold Brick Mining District of Colorado - Important insights into the history of the Gold Brick Mining District, as well as its local geography and economic geology. Also included are the histories and locations of historic mines in this important Colorado Mining District, including the Cortland, Carter, Raymond, Gold Links, Sacramento, Bassick, Sandy Hook, Chronicle, Grand Prize, Chloride, Granite Mountain, Lucille, Gray Mountain, Hilltop, Maggie Mitchell, Silver Islet, Revenue, Roosevelt, Carbonate King and others. In addition to hardrock mining, are also included are details on gold placer mining in this portion of Colorado. **8.5" X 11", 140 ppgs, Retail Price: $12.99**

Washington Mining Books

The Republic Mining District of Washington - Unavailable since 1910, this important publication was originally published by the Washington Geologic Survey and has been unavailable for a century. Topics include the geology, rock formations and the formation of ore deposits in this important mining area of Washington State. Also included are hard to find details on the geology, history and locations of dozens of mines in the area. Some of the mines featured include the New Republic Mine, Ben Hur, Morning Glory, the South Republic Mine, Quilp, Surprise, Black Tail, Lone Pine, San Poil, Mountain Lion, Tom Thumb, Elcaliph and many others. **8.5" X 11", 94 ppgs, Retail Price: $10.99**

The Myers Creek and Nighthawk Mining Districts of Washington - Unavailable since 1911, this important publication was originally published by the Washington Geologic Survey and has been unavailable for a century. Topics include the geology, rock formations and the formation of ore deposits in these important mining areas of Washington State. Also included are hard to find details on the geology, history and locations of dozens of mines in the area. Some of the mines featured include the Grant Mine, Monterey, Nip and Tuck, Myers Creek, Number Nine, Neutral, Rainbow, Aztec, Crystal Butte, Apex, Butcher Boy, Molson, Mad River, Olentangy, Delate, Kelsey, Golden Chariot, Okanogan, Ohio, Forty-Ninth Parallel, Nighthawk, Favorite, Little Chopaka, Summit, Number One, California, Peerless, Caaba, Prize Group, Ruby, Mountain Sheep, Golden Zone, Rich Bar, Similkameen, Kimberly, Triune, Hiawatha, Trinity, Hornsilver, Maquae, Bellevue, Bullfrog, Palmer Lake, Ivanhoe, Copper World and many others.
 8.5″ X 11″, 136 ppgs, Retail Price: $12.99

The Blewett Mining District of Washington - Unavailable since 1911, this important publication was originally published by the Washington Geologic Survey and has been unavailable for a century. Topics include the geology, rock formations and the formation of ore deposits in this important mining area of Washington State. Also included are hard to find details on the geology, history and locations of dozens of mines in the area. Some of the mines featured include the Washington Meteor, Alta Vista, Pole Pick, Blinn, North Star, Golden Eagle, Tip Top, Wilder, Golden Guinea, Lucky Queen, Blue Bell, Prospect, Homestake, Lone Rock, Johnson, and others. **8.5″ X 11″, 134 ppgs, Retail Price: $12.99**

Silver Mining In Washington - Unavailable since 1955, this important publication was originally published by the Washington Geologic Survey. Featured are the hard to find locations and details pertaining to Washington's silver mines. **8.5″ X 11″, 180 ppgs, Retail Price: $15.99**

The Mines of Snohomish County Washington - Unavailable since 1942, this important publication was originally published by the Washington Geologic Survey and has been unavailable for seventy years. Featured are details on a large number of gold, silver, copper, lead and other metallic mineral mines. Included are the locations of each historic mine, along with information on the commodity produced. **8.5″ X 11″, 98 ppgs, Retail Price: $10.99**

The Mines of Chelan County Washington - Unavailable since 1943, this important publication was originally published by the Washington Geologic Survey and has been unavailable for seventy years. Featured are details on a large number of gold, silver, copper, lead and other metallic mineral mines. Included are the locations of each historic mine, along with information on the commodity. **8.5″ X 11″, 88 ppgs, Retail Price: $9.99**

Metal Mines of Washington - Unavailable since 1921, this important publication was originally published by the Washington Geologic Survey and has been unavailable for nearly ninety years. Widely considered a masterpiece on the Washington Mining Industry, "Metal Mines of Washington" sheds light on the important details of Washington's early mining years. Featured are details on hundreds of gold, silver, copper, lead and other metallic mineral mines. Included are hard to find details on the mineral resources of this state, as well as the locations of historic mines. Lavishly illustrated with maps and historic photos and complete with a glossary to explain any technical terms found in the text, this is one of the most important works on mining in the State of Washington. No prospector or miner should be without it if they are interested in mining in Washington. **8.5″ X 11″, 396 ppgs, Retail Price: $24.99**

Gem Stones In Washington - Unavailable since 1949, this important publication was originally published by the Washington Geologic Survey and has been unavailable since first published. Included are details on where to find naturally occurring gem stones in the State of Washington, including quartz crystal, amethyst, smoky quartz, milky quartz, agates, bloodstone, carnelian, chert, flint, jasper, onyx, petrified wood, opal, fire opal, hyalite and others. **8.5″ X 11″, 54 ppgs, Retail Price: $8.99**

The Covada Mining District of Washington - Unavailable since 1913, this important publication was originally published by the Washington Geologic Survey and has been unavailable for a century. Topics include the geology, rock formations and the formation of ore deposits in this important mining area of Washington State. Also included are hard to find details on the geology, history and locations of dozens of mines in the area. Some of the mines featured include the Admiral, Advance, Algonkian, Big Bug, Big Chief, Big Joker, Black Hawk, Black Tail, Black Thorn, Captain, Cherokee Strip, Colorado, Dan Patch, Dead Shot, Etta, Good Ore, Greasy Run, Great Scott, Idora, IXL, Jay Bird, Kentucky Bell, King Solomon, Laurel, Laura S, Little Jay, Meteor, Neglected, Northern Light, Old Nell, Plymouth Rock, Polaris, Quandary, Reserve, Shoo Fly, Silver Plume, Three Pines, Vernie, White Rose and dozens of others. **8.5″ X 11″, 114 ppgs, Retail Price: $10.99**

The Index Mining District of Washington - Unavailable since 1912, this important publication was originally published by the Washington Geologic Survey and has been unavailable for a century. Topics include the geology, rock formations and the formation of ore deposits in this important mining area of Washington State. Also included are hard to find details on the geology, history and locations of dozens of mines in the area. Some of the mines featured include the Sunset, Non-Pareil, Ethel Consolidated, Kittaning, Merchant, Homestead, Co-operative, Lost Creek, Uncle Sam, Calumet, Florence-Rae, Bitter Creek, Index Peacock, Gunn Peak, Helena, North Star, Buckeye. Copper Bell, Red Cross and others. **8.5″ X 11″, 114 ppgs, Retail Price: $11.99**

Mining & Mineral Resources of Stevens County Washington - Unavailable since 1920, this important publication was originally published by the Washington Geologic Survey and has been unavailable for a century. Topics include the geology, rock formations and the formation of ore deposits in these important mining areas of Washington State. Also included are hard to find details on the geology, history and locations of hundreds of mines in the area. **8.5" X 11", 372 ppgs, Retail Price: $24.99**

The Mines and Geology of the Loomis Quadrangle Okanogan County, Washington - Unavailable since 1972, this important publication was originally published by the Washington Geologic Survey and has been unavailable for a century. Topics include the geology, rock formations and the formation of ore deposits in this important mining area of Washington State. Also included are hard to find details on the geology, history and locations of dozens of gold, copper, silver and other mines in the area. **8.5" X 11", 150 ppgs, Retail Price: $12.99**

The Conconully Mining District of Okanogan County Washington - Unavailable since 1973, this important publication was originally published by the Washington Geologic Survey and has been unavailable for a century. Topics include the geology, rock formations and the formation of ore deposits in this important mining area of Washington State, which also includes Salmon Creek, Blue Lake and Galena. Also included are hard to find details on the geology, mining history and locations of dozens of mines in the area. Some of the mines include Arlington, Fourth of July, Sonny Boy, First Thought, Last Chance, War Eagle-Peacock, Wheeler, Mohawk, Lone Star, Woo Loo Moo Loo, Keystone, Hughes, Plant-Callahan, Johnny Boy, Leuena, Gubser, John Arthur, Tough Nut, Homestake, Key and many others **8.5" X 11", 68 ppgs, Retail Price: $8.99**

Wyoming Mining Books

Mining in the Laramie Basin of Wyoming - Unavailable since 1909, this publication was originally compiled by the United States Department of Interior. Also included are insights into the mineralization and other characteristics of this important mining region, especially in regards to coal, limestone, gypsum, bentonite clay, cement, sand, clay and copper. **8.5" X 11", 104 ppgs, Retail Price: $11.99**

New Mexico Mining Books

The Mogollon Mining District of New Mexico - Unavailable since 1927, this important publication was originally published by the US Department of Interior and has been unavailable for 80 years. Topics include the geology, rock formations and the formation of ore deposits in this important mining area in New Mexico. Of particular focus is information on the history and production of the ore deposits in this area, their form and structure, vein filling, their paragenesis, origins and ore shoots, as well as oxidation and supergene enrichment. Also included are hard to find details, including the descriptions and locations of numerous gold, silver and other types of mines, including the Eureka, Pacific, South Alpine, Great Western, Enterprise, Buffalo, Mountain View, Floride, Gold Dust, Last Chance, Deadwood, Confidence, Maud S., Deep Down, Little Fanney, Trilby, Johnson, Alberta, Comet, Golden Eagle, Cooney, Queen, the Iron Crown, Eberle, Clifton, Andrew Jackson mine, Mascot and others. **8.5" X 11", 144 ppgs, Retail Price: $12.99**

The Percha Mining District of Kingston New Mexico - Unavailable since 1883, this important publication was originally published by the Kingston Tribune and has been unavailable for over one hundred and thirty five years. Having been written during the earliest years of gold and silver mining in the Percha Mining District, unlike other books on the subject, this work offers the unique perspective of having actually been written while the early mining history of this area was still being made. In fact, the work was written so early in the development of this area that many of the notable mines in the Percha District were less than a few years old and were still being operated by their original discoverers with the same enthusiasm as when they were first located. Included are hard to find details on the very earliest gold and silver mines of this important mining district near Kingston in Sierra County, New Mexico. **8.5" X 11", 68 ppgs, Retail Price: $9.99**

East Coast Mining Books

The Gold Fields of the Southern Appalachians - Unavailable since 1895, this important publication was originally published by the US Department of Interior and has been unavailable for nearly 120 years. Topics include the geology, rock formations and the formation of ore deposits in this important mining area of the American South. Of particular focus is information on the history and statistics of the ore deposits in this area, their form and structure and veins. Also included are details on the placer gold deposits of the region. The gold fields of the Georgian Belt, Carolinian Belt and the South Mountain Mining District of North Carolina are all treated in descriptive detail. Included are hard to find details, including the descriptions and locations of numerous gold mines in Georgia, North Carolina and elsewhere in the American South. Also included are details on the gold belts of the British Maritime Provinces and the Green Mountains. **8.5" X 11", 104 ppgs, Retail Price: $9.99**

Gold Rush Tales Series

Millions in Siskiyou County Gold - In this first volume of the "Gold Rush Tales" series, leading mining historian and editor Kerby Jackson, introduces us to the story of how millions of dollars worth of gold was discovered in Siskiyou County during the California Gold Rush. Lavishly illustrated with photos from the 19th Century, this hard to find information was first published in 1897 and sheds important light onto the gold rush era in Siskiyou County, California and the experiences of the men who dug for the gold and actually found it. **8.5" X 11", 82 ppgs, Retail Price: $9.99**

The California Rand in the Days of '49 - In this second volume of the "Gold Rush Tales" series, leading mining historian and editor Kerby Jackson, introduces us to four tales from the California Gold Rush. Lavishly illustrated with photos from the 19th Century, this hard to find information was first published in 1890's and includes the stories of "California's Rand", details about Chinese miners, how one early miner named Baker struck it rich and also the story of Alphonzo Bowers, who invented the first hydraulic gold dredge. **8.5" X 11", 54 ppgs, Retail Price: $9.99**

More Mining Books

Prospecting and Developing A Small Mine - Topics covered include the classification of varying ores, how to take a proper ore sample, the proper reduction of ore samples, alluvial sampling, how to understand geology as it is applied to prospecting and mining, prospecting procedures, methods of ore treatment, the application of drilling and blasting in a small mine and other topics that the small scale miner will find of benefit. **8.5" X 11", 112 ppgs, Retail Price: $11.99**

Timbering For Small Underground Mines - Topics covered include the selection of caps and posts, the treatment of mine timbers, how to install mine timbers, repairing damaged timbers, use of drift supports, headboards, squeeze sets, ore chute construction, mine cribbing, square set timbering methods, the use of steel and concrete sets and other topics that the small underground miner will find of benefit. This volume also includes twenty eight illustrations depicting the proper construction of mine timbering and support systems that greatly enhance the practical usability of the information contained in this small book. **8.5" X 11", 88 ppgs. Retail Price: $10.99**

Timbering and Mining - A classic mining publication on Hard Rock Mining by W.H. Storms. Unavailable since 1909, this rare publication provides an in depth look at American methods of underground mine timbering and mining methods. Topics include the selection and preservation of mine timbers, drifting and drift sets, driving in running ground, structural steel in mine workings, timbering drifts in gravel mines, timbering methods for driving shafts, positioning drill holes in shafts, timbering stations at shafts, drainage, mining large ore bodies by means of open cuts or by the "Glory Hole" system, stoping out ore in flat or low lying veins, use of the "Caving System", stoping in swelling ground, how to stope out large ore bodies, Square Set timbering on the Comstock and its modifications by California miners, the construction of ore chutes, stoping ore bodies by use of the "Block System", how to work dangerous ground, information on the "Delprat System" of stoping without mine timbers, construction and use of headframes and much more. This volume provides a reference into not only practical methods of mining and timbering that may be employed in narrow vein mining by small miners today, but also rare insights into how mines were being worked at the turn of the 19th Century. **8.5" X 11", 288 ppgs. Retail Price: $24.99**

A Study of Ore Deposits For The Practical Miner - Mining historian Kerby Jackson introduces us to a classic mining publication on ore deposits by J.P. Wallace. First published in 1908, it has been unavailable for over a century. Included are important insights into the properties of minerals and their identification, on the occurrence and origin of gold, on gold alloys, insights into gold bearing sulfides such as pyrites and arsenopyrites, on gold bearing vanadium, gold and silver tellurides, lead and mercury tellurides, on silver ores, platinum and iridium, mercury ores, copper ores, lead ores, zinc ores, iron ores, chromium ores, manganese ores, nickel ores, tin ores, tungsten ores and others. Also included are facts regarding rock forming minerals, their composition and occurrences, on igneous, sedimentary, metamorphic and intrusive rocks, as well as how they are geologically disturbed by dikes, flows and faults, as well as the effects of these geologic actions and why they are important to the miner. Written specifically with the common miner and prospector in mind, the book will help to unlock the earth's hidden wealth for you and is written in a simple and concise language that anyone can understand. **8.5" X 11", 366 ppgs. Retail Price: $24.99**

Mine Drainage - Unavailable since 1896, this rare publication provides an in depth look at American methods of underground mine drainage and mining pump systems. This volume provides a reference into not only practical methods of mining drainage that may be employed in narrow vein mining by small miners today, but also rare insights into how mines were being worked at the turn of the 19th Century. **8.5" X 11", 218 ppgs. Retail Price: $24.99**

Fire Assaying Gold, Silver and Lead Ores - Unavailable since 1907, this important publication was originally published by the Mining and Scientific Press and was designed to introduce miners and prospectors of gold, silver and lead to the art of fire assaying. Topics include the fire assaying of ores and products containing gold, silver and lead; the sampling and preparation of ore for an assay; care of the assay office, assay furnaces; crucibles and scorifiers; assay balances; metallic ores; scorification assays; cupelling; parting' crucible assays, the roasting of ores and more. This classic provides a time honored method of assaying put forward in a clear, concise and easy to understand language that will make it a benefit to even beginners. **8.5" X 11", 96 ppgs. Retail Price: $11.99**

Methods of Mine Timbering - Originally published in 1896, this important publication on mining engineering has not been available for nearly a century. Included are rare insights into historical methods of timbering structural support that were used in underground metal mines during the California that still have a practical application for the small scale hardrock miner of today. **8.5" X 11", 94 ppgs. Retail Price: $10.99**

The Enrichment of Copper Sulfide Ores - First published in 1913, it has been unavailable for over a century. Topics include the definition and types of ore enrichment, the oxidation of copper ores, the precipitation of metallic sulfides. Also included are the results of dozens of lab experiments pertaining to the enrichment of sulfide ores that will be of interest to the practical hard rock mine operator in his efforts to release the metallic bounty from his mine's ore. **8.5" X 11", 92 ppgs. Retail Price: $9.99**

A Study of Magmatic Sulfide Ores - Unavailable since 1914, this rare publication provides an in depth look at magmatic sulfide ores. Some of the topics included are the definition and classification of magmatic ores, descriptions of some magmatic sulfide ore deposits known at the time of publication including copper and nickel bearing pyrrohitic ore bodies, chalcopyrite-bornite deposits, pyritic deposits, magnetite-ileminite deposits, chromite deposits and magmatic iron ore deposits. Also included are details on how to recognize these types of ore deposits while prospecting for valuable hardrock minerals. **8.5" X 11", 138 ppgs. Retail Price: $11.99**

The Cyanide Process of Gold Recovery - Unavailable since 1894 and released under the name "The Cyanide Process: Its Practical Application and Economical Results", this rare publication provides an in depth look at the early use of cyanide leaching for gold recovery from hardrock mine ores. This volume provides a reference into the early development and use of cyanide leaching to recover gold. **8.5" X 11", 162 ppgs. Retail Price: $14.99**

California Gold Milling Practices - Unavailable since 1895 and released under the name "California Gold Practices", this rare publication provides an in depth look at early methods of milling used to reduce gold ores in California during the late 19th century. This volume provides a reference into the early development and use of milling equipment during the earliest years of the California Gold Rush up to the age of the Industrial Revolution. Much of the information still applies today and will be of use to small scale miners engaging in hardrock mining. **8.5" X 11", 104 ppgs. Retail Price: $10.99**

www.ingramcontent.com/pod-product-compliance
Lightning Source LLC
Chambersburg PA
CBHW080819180526
45168CB00006B/2503

* 9 781506 172668 *